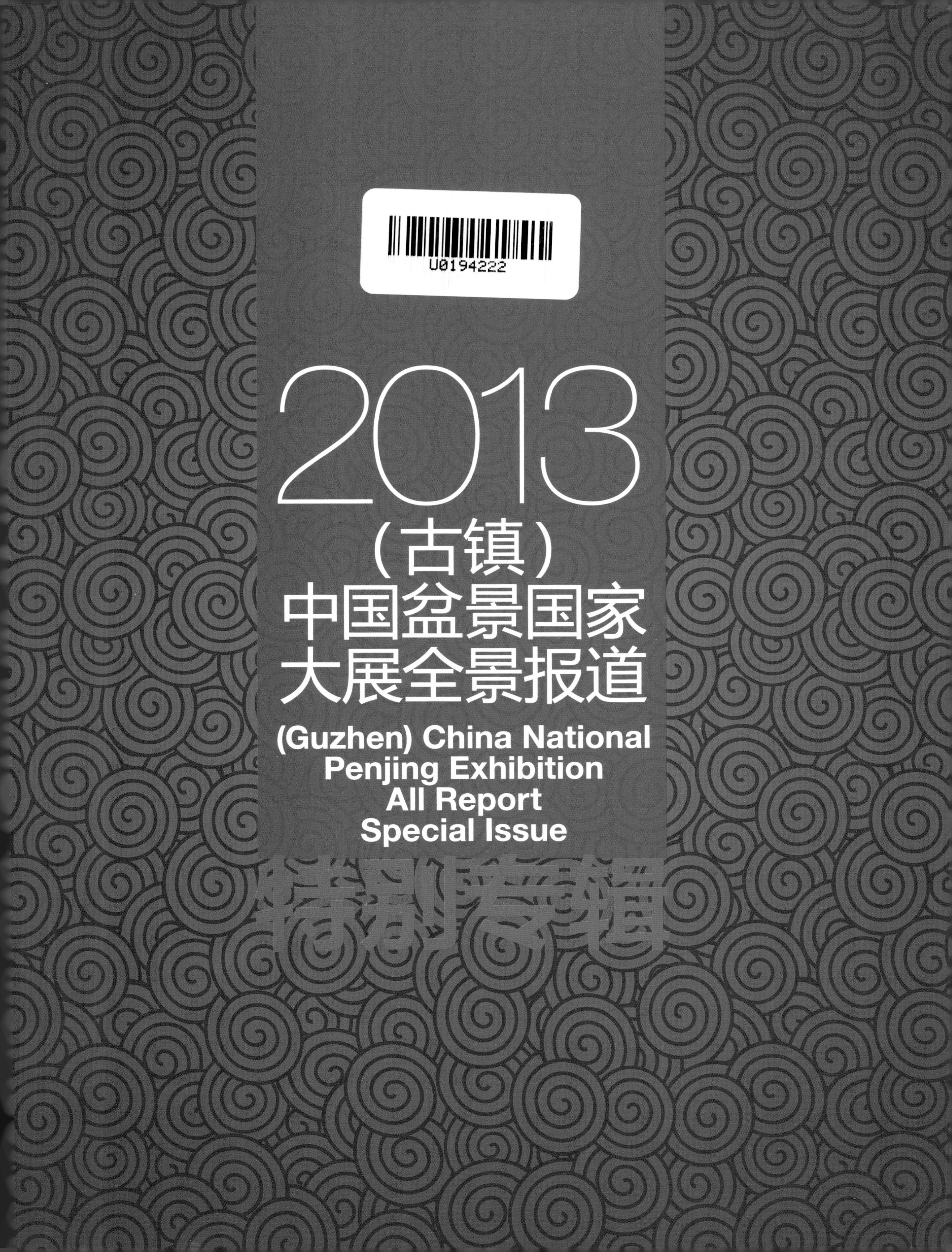
U0194222
2013
(古镇)
中国盆景国家
大展全景报道
(Guzhen) China National
Penjing Exhibition
All Report
Special Issue
特别专辑

中国盆景赏石

2013(Guzhen)China National Penjing Exhibition All Report Special Issue
2013（古镇）中国盆景国家大展全景报道特别专辑

中国林业出版社 China Forestry Publishing House

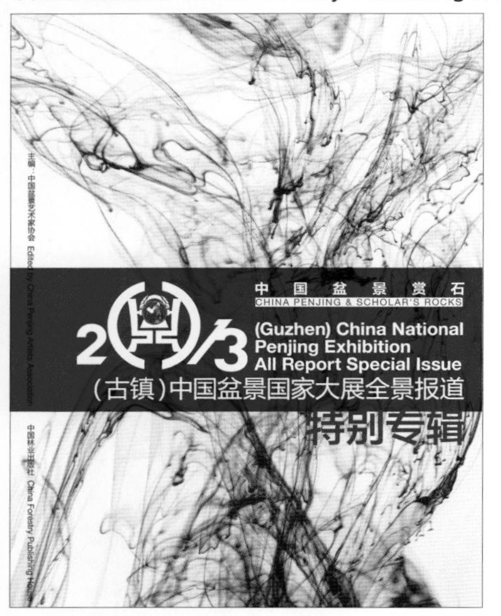

向世界一流水准努力的
——中文高端盆景媒体

《中国盆景赏石》

世界上第一本全球发行的中文大型盆景媒体
向全球推广中国盆景文化的传媒大使
为中文盆景出版业带来全新行业标准

本期摄影除署名外均由纪武军拍摄

图书在版编目（CIP）数据

中国盆景赏石 . 2013.（古镇）中国盆景国家大展全景报道特别专辑 / 中国盆景艺术家协会主编 . -- 北京 : 中国林业出版社，2014.1
ISBN 978-7-5038-7369-0
Ⅰ.①中… Ⅱ.①中… Ⅲ.①盆景－观赏园艺－中国－文集②观赏型－石－中国－文集 Ⅳ.① S688.1-53 ② TS933-53
中国版本图书馆 CIP 数据核字 (2014) 第 012034 号

责任编辑：何增明 张华
出 版：中国林业出版社
E-mail:shula5@163.com
电话：(010) 83286967
社 址：北京西城区德内大街刘海胡同 7 号
邮编：100009
发 行：中国林业出版社
印 刷：北京利丰雅高长城印刷有限公司
开 本：230mm×300mm
版 次：2014 年 1 月第 1 版
印 次：2014 年 1 月第 1 次
印 张：11.5
字 数：300 千字
定 价：48.00 元

主办、出品、编辑：  中国盆景艺术家协会
E-mail: penjingchina@aliyun.com.cn
Sponsor/Produce/Edit: China Penjing Artists Association

创办人、总出版人、总编辑、视觉总监、摄影： 苏放
Founder，Publisher，Editor-in-Chief，Visual Director，Photographer: Su Fang
电子邮件： E-mail : 1440372565@qq.com

编辑报道热线： 010-58693878（每周一至五：上午10：00-下午6：30）
News Report Hotline: 010-58693878 (10 : 00a.m to 6 : 30p.m, Monday to Friday)
传真Fax: 010-58693878
投稿邮箱Contribution E-mail: CPSR@foxmail.com
会员订阅及协会事务咨询热线： 010-58690358（每周一至五：上午10：00-下午6：30）
Subscribe and Consulting Hotline: 010-58690358 (10 : 00a.m to 6 : 30p.m, Monday to Friday)
通信地址： 北京市朝阳区建外SOHO16号楼1615室 **邮编：** 100022
Address: JianWai SOHO Building 16 Room 1615, Beijing ChaoYang District, 100022 China

法律顾问： 赵煜 **Legal Counsel:** Zhao Yu

制版印刷： 北京利丰雅高长城印刷有限公司
读者凡发现本书有掉页、残页、装订有误等印刷质量问题，请直接邮寄到以下地址，印刷厂将负责退换：北京市通州区中关村科技园通州光机电一体化产业基地政府路2号 邮编101111
联系人王莉，电话：010-59011332。

点评 Comments

06 "双雄竞秀" 九里香 *Murraya exotica* 陈伟藏品——2013 中国鼎国家大展首席大奖 文：罗传忠、王选民、徐昊
"Heroic Duo Contend for Beauty". Collector: Chen Wei—2013 China Ding Grand Prize of National Exhibition
Author: Luo Chuanzhong, Wang Xuanmin, Xu Hao

盆景中国 Penjing China

10 中国鼎 向下一个时代进军的号角——中国鼎 2013（古镇）中国盆景国家大展全景报道 报道：CP
China Ding Horn for Marching to the Next Era—China Ding 2013(Guzhen)China National Penjing Exhibition All Report Reporter: CP

点评 Comments

20 "雄风" 刺柏 *Juniperus formosana* 鲍世骐藏品——2013 中国盆景国家大展奖 文：王选民
"Heroic Spirit". Chinese Juniper. Collector: Bao Shiqi—2013 China National Penjing Exhibition Award Author: Wang Xuanmin

22 "奇劲唱风" 赤松 *Pinus densiflora* 曹志振藏品——2013 中国盆景国家大展奖 文：徐昊
"The Wind Blowing with Powerful Vigor". Japanese Red Pine. Collector: Cao Zhizhen—2013 China National Penjing Exhibition Award Author: Xu Hao

24 "疑是枝头蝶恋花" 簕杜鹃 *Bougaivillea spectabillis* 吴成发藏品——2013 中国盆景国家大展奖 文：李奕祺
"Just Like the Circling Butterflies above the Tree Top Staying for the Flowers". Paper Flower. Height: 120cm. Collector: Ng Shingfat—2013 China Ding China National Penjing Exhibition Award Author: Li Yiqi

26 "龙飞凤舞" 榕树 *Ficus microcarpa* 李正银藏品——2013 中国盆景国家大展奖 文：李奕祺
"Like Dragons Flying and Phoenixes Dancing". Chinese Banyan. Collector: Li Zhengyin—2013 China National Penjing Exhibition Award Author: Li Yiqi

28 九里香 *Murraya exotica* 罗崇辉藏品——2013 中国盆景国家大展奖 文：张志刚
Jasminorange. Collector: Luo Chonghui—2013 China National Penjing Exhibition Award Author: Zhang Zhigang

30 真柏 *Juniperus chinensis* var. *sargentii* 陈国健藏品——2013 中国盆景国家大展奖 文：罗传忠
Sargent Savin. Collector: Chen Guojian —2013 China National Penjing Exhibition Award Author: Luo Chuanzhong

32 "盛世" 五针松 *Pinus parviflora* 杨贵生藏品——2013 中国盆景国家大展奖 文：徐昊
"Flourishing Age". Japanese White Pine. Collector: Yang Guisheng—2013 China National Penjing Exhibition Award Author: Xu Hao

34 "天娇" 真柏 *Juniperus chinensis* var. *sargentii* 陈伟藏品——2013 中国盆景国家大展奖 文：张志刚
"Delicate and Charming Girl of Heaven". Sargent Savin. Collector: Chen Wei—2013 China National Penjing Exhibition Award Author: Zhang Zhigang

论坛中国 Forum China

36 畅谈中国鼎——2013（古镇）中国盆景国家大展印象
The Impressions of China Ding—2013(Guzhen) China National Penjing Exhibition

52 中西碰撞 艳惊寰宇——国际著名盆景人眼中的"中国鼎"
Penjing Culture Clash Between China and the West Surprise the World—Famous International Penjing People's Views on "China Ding"

话题 Issue

72 中国鼎——2013（古镇）中国盆景国家大展展品选拔筹备工作回顾 文：鲍世骐
China Ding—2013(Guzhen) China National Penjing Exhibition Exhibits Selection and Preparatory Work Review Author: Bao Shiqi

76 用爱编织我们的中国盆景艺术家协会——国际盆栽大师梁悦美教授访谈
受访人 / 供图：梁悦美 访谈人：CP
Woven Our China Penjing Artists Association with Love
—An Interview with Professor Amy Liang, the International Bonsai Master
Interviewee / Photo Provider : Amy Liang Interviewer: CP

80 中国盆景国家大展观后感 文：李正银、罗传忠
Impressions of China National Penjing Exhibition
Authors: Li Zhengyin & Luo Chuanzhong

中国现场 On-the-Spot

82 展示松树造型艺术的正能量——樊顺利大师在 2013（古镇）中国盆景国家大展上的现场制作表演 制作：樊顺利 文：胡光生
Show the Positive Energy of Pines' Plastic Arts—Master Fan Shunli's Demonstration Performance of 2013(Guzhen) China National Penjing Exhibition
Processor: Fan Shunli Author: Hu Guangsheng

88 随形就势，因材施艺——2013（古镇）中国盆景国家大展现场制作表演 文：徐昊
Making due to the Material—Demonstration Performance of 2013 (Guzhen) China National Penjing Exhibition Author: Xu Hao

会员展获奖作品专栏 The Column of Winning Works

94 2013 中国盆景会员展年度大奖作品欣赏
The Annual Award Works Appreciation of 2013 China Penjing Member Exhibition

95 2013 中国盆景会员展金奖作品欣赏
Gold Award Works Appreciation of 2013 China Penjing Member Exhibition

96 2013 中国盆景会员展银奖作品欣赏
Silver Award Works Appreciation of 2013 China Penjing Member Exhibition

98 2013 中国盆景会员展铜奖作品欣赏
Bronze Award Works Appreciation of 2013 China Penjing Member Exhibition

102 2013 中国盆景艺术家协会会员盆景精品展评比计分表
Assessment Scoring Form of China Penjing Member Exhibition of China Penjing Artists Association

古盆中国 China Ancient Pot

105 清早期 炉钧釉圆飘口圈足盆 直径 38.5cm 高 20.5cm 杨贵生藏品
Early Qing Dynasty Robin's Egg with Round Overhanging-Edge Pot. Diameter: 38.5cm, Height: 20.5cm. Collector: Yang Guisheng

106 清早期 乌泥圆八卦纹兽脸足盆 直径 39.5cm 宽 11.3 cm 申洪良藏品
Early Qing Dynasty Dark Clay with Trigram Pattern and Feast Face Leg Pot. Diameter: 39.5cm, Width: 11.3cm. Collector: Shen Hongliang

107 清早期 乌泥树桩型盆 长 38cm 宽 32cm 高 15.5cm 杨贵生藏品
Early Qing Dynasty Dark Clay in Tree Stump Shape Pot. Length: 38cm, Width: 32cm, Height: 15.5cm. Collector: Yang Guisheng

108 清初 乌泥圆鼓钉盆 直径 37cm 高 8 cm 申洪良藏品
Early Qing Dynasty Dark Clay with Round Nail Shape Pot. Diameter: 37cm, Height: 8cm. Collector: Shen Honglian g

109 清初 紫泥长方飘口底线墨彩盆 长 54cm 宽 33cm 高 29.5cm 杨贵生藏品
Early Qing Dynasty Purple Clay with Chinese Painting and Overhanging-Edge Rectangular Pot. Length: 54cm, Width: 33cm, Height: 29.5cm. Collector: Yang Guisheng

110 清初 乌泥海棠飘口上下带线盆 长 48cm 宽 39.5cm 高 26cm 申洪良藏品
Early Qing Dynasty Dark Clay with Two Stripe and Crabapple Shape Overhanging-Edge Pot. Length: 48cm, Width: 39.5cm, Height: 26cm. Collector: Shen Hongliang

111 明末 乌泥长方直壁上下带线盆 长 45.5cm 宽 29cm 高 17.5cm 杨贵生藏品
Late-Ming Dynasty Dark Clay with Two Lines Rectangular Pot. Length: 45.5cm, Width: 29cm, Height: 17.5cm. Collector: Yang Guisheng

会员展获奖作品专栏 The Column of Winning Works

112 2013 中国盆景会员展之古盆欣赏
Ancient Pot Appreciation of 2013 China Penjing Member Exhibition

话题 Issue

116 可圈可点——我看中国盆景国家大展 文：刘洪
Commendable—I see China National Penjing Exhibition Author: Liu Hong

118 浅谈中国鼎 2013（古镇）中国盆景国家大展 文：曾尔恩
Talk About China Ding 2013(Guzhen) China National Penjing Exhibition
Author：Zeng Er'en

120 初窥中国鼎——2013（古镇）中国盆景国家大展和华风展之浅见 文：罗瑞本
View on the Differences Between China Ding—2013 (Guzhen) China National Penjing Exhibition and Huafeng Bonsai Exhibition Author：Luo Ruiben

盆景中国 Penjing China

126 2013 中国鼎之旅——2013（古镇）中国盆景国家大展后中外嘉宾参观团访问之中国唐苑篇
2013 China Ding Trip—Foreign Guests' Delegation Visited Chinese Tangyuan Garden after 2013 (Guzhen) China National Penjing Exhibition

138 2013 中国鼎之旅——2013（古镇）中国盆景国家大展后中外嘉宾参观访问之趣怡园篇
2013 China Ding Trip—Sino - Foreign Guests Visited the Quyi Garden after 2013 (Gu Zhen) China National Penjing Exhibition

148 2013 中国鼎之旅——2013（古镇）中国盆景国家大展后中外嘉宾参观访问之真趣园篇
2013 China Ding Trip—Foreign Guests' Delegation Visited Zhenqu Garden after 2013 (Guzhen) China National Penjing Exhibition

赏石中国 China Scholar' s Rocks

158 "琼楼玉宇" 玛瑙 长 50cm 宽 40cm 高 39cm 马建中藏品
"Richly Palace". Agate. Length: 50cm, Width: 40cm, Height: 39cm.
Collector: Ma Jianzhong

159 "红河艳" 大化石 长 27cm 宽 31cm 高 18cm 李正银藏品
" Colourful Red Rival". Macrofossil. Length: 27cm, Width: 31cm, Height: 18cm.
Collector: Li Zhengyin

160 2013 中国盆景会员展之赏石欣赏
Scholar's Rocks Appreciation of 2013 China Penjing Member Exhibition

国际盆景世界 Penjing International

164 韩国盆栽园集锦
Introduction of Korean Bonsai Garden

盆景中国 Penjing China

174 中国鼎 2013（古镇）中国盆景国家大展展场花絮
China Ding 2013(Guzhen)China National Penjing Exhibition
A Complete Collection of the Exhibition Scene

178 2013（古镇）中国盆景国家大展筹备会议与总结会议实录
The Records of 2013(Guzhen)China National Penjing Exhibition Preparatory Meetings and Wrap-up Meeting

“双雄竞秀”九里香 *Murraya exotica* 陈伟藏品
—— 2013中国鼎国家大展首席大奖

“Heroic Duo Contend for Beauty”. Collector: Chen Wei
—2013 China Ding Chief Prize of National Exhibition

文：王选民 Author: Wang Xuanmin

“双雄竞秀”九里香 *Murraya exotica* 高92cm 宽168cm 陈伟藏品 苏放摄影
"Heroic Duo Contend for Beauty". Jasminorange. Height: 92cm, Width: 168cm. Collector: Chen Wei, Photographer: Su Fang

点评 Comments

九里香是南方区域性树种，其树性本身具有许多特征性优点，故在南方杂木类当中颇有人气。但能正确把握九里香的特点创造出理想的作品的人，确实有一定的水平。

这棵树就素材特点上而言，可谓百里挑一。造型上已有十年之多，枝干培养功深可见。尽管在素材特点的取势定位和枝法应用方面尚需要完善，但树的自然形象和九里香所固有的天性特征还是能打动人心。其作品的神采和枝叶之间的自然气息足以让人入神，玩味无穷。

Murraya is the southern regional species. The tree itself has many characteristic advantages, so it's a kind of tree that is very popular among the southern miscellaneous trees. But there's a certain degree of difficulty to grasp its characteristic correctly to create the ideal work.

This tree is one in a hundred in terms of the material characteristics. It can be seen the branches and stem trained well, the shape of the tree has more than ten years. It still needs to be improved in term of the potential position of material characteristics and the application of branch techniques, but it can go directly to the heart from the aspect of tree's nature images and the inherent characteristics of nature. People are deeply enthralled, pondering infinity by the work's presence and the natural atmosphere between the branches and leaves.

"双雄竞秀" 九里香 *Murraya exotica* 陈伟藏品——2013中国鼎国家大展首席大奖

"Heroic Duo Contend for Beauty". Jasminorange. Collector: Chen Wei — 2013 China Ding Chief Prize of National Exhibition

文：徐昊 Author: Xu Hao

经常有人说："盆景是自然的浓缩与再现"。对于一般所看到的盆景或盆栽而言，可能就是那么回事，而对于一件盆景艺术作品来说，这话就有失偏颇。

中国盆景的审美，不仅观其外相，更注重看其内真。所谓外相，便是树桩经过裁剪、造型，养护完善后的动止形态；所谓内真，则是造型布势的气度和蕴含其中的人文精神。一件优秀的中国盆景艺术作品，绝不是"再现"古树名木的自然形态或一隅优美的山水风景，而是胸中丘壑，是心化了的景观，是"再造"而不是"再现"。再造自然为心化的自然，其创作发之于心，行之于迹，立之于象，现之于景，于景生境，境生于象外，即景外之情、之境。陈伟先生的九里香盆景"双雄竞秀"，便是这样一件优秀的盆景艺术作品，因此在首届中国盆景国家大展的100件盆景精品中脱颖而出，获得大展首奖——"中国鼎"奖。

作品根盘稳健，根爪或匍匐起伏，或腾跃曲折，紧扎地面，充满稳定的力量感。主干呈左斜之势，线条曲折优美，干肌隆张劲健。主干下部粗壮，看似由于树龄的古老和岁月的涤荡，使之渐渐中空，且一分为二，形成自然的双干树相。双干之间尚有舍利相对，平添了作品的岁月沧桑和历史印记，增加了作品深邃的时空内涵。作者根据树桩素材的特点，取低位出枝，以向左的主枝凌空展宕，来加强作品的奇险之势和空间之美；以副干粗壮而向右舒展的树枝来稳定重心，使作品于奇险中复归平正稳定。树枝的线条曲折起伏，遒劲有力，整体的枝势或展宕或飘逸，或平舒或上扬，极富变化之美。错落变化的枝线使作品的结构显得充满跌宕起伏的矛盾和抗争，而作者巧妙地驾驭树势和主线条的走向，使作品于矛盾中达到大势的协调和统一，也因此而增加了作品可赏的内容，深化了作品的内涵。

作品的外相，是雄浑博大的气势和飘逸儒雅的姿态，满树新叶初萌，让人如沐春风，一番和煦明媚的景象。葱倩优雅的内在，却是百折不回的傲然风骨，风骨里，既是自然沧桑的经历和抗争，也是人类自我精神的体现。

People usually say: "Penjing is the concentration and reappearance of the nature". It may be right for some ordinary Penjing or Bonsai but for a Penjing artworks, this is not quite right.

The beauty-appreciation of Chinese Penjing is not just about the appearance but more important in its inside truth. The so called appearance means the static gesture of tree stump after trimming, modeling, maintaining and improvement; and the so called truth means the attitude of modeling layout and the humanistic spirit contained inside. An excellent artwork of China Penjing art is not just a reappearance of the natural gesture of the ancient tree or a beautiful corner of landscape scenery, but becomes mountains and rivers in one's heart, a view in the mind. It is not to "reappear", but to "rebuild". The rebuilt nature is a nature with heart. The creation begins with heart, forms from traces, rises in images and appears in a view, which creates scenery. The scenery rises out of the images, which means the motion and scenery are beyond the view. The *Murraya exotica* Penjing "Heroic Duo Contend for Beauty" of Mr. Chen Wei from Fujian is a great Penjing work of such kind. Thus it stands out from the 100 Penjing collections in the National Exhibition of China Penjing and won the Chief Prize – "China Ding".

The work is steady at the stump with root claws or creeping ups and downs, or vaulting and twisting, and it is tightly rooted to the ground with firm strength. The trunk leans to the left with beautiful and twisted lines, as the developed and powerful muscle. The lower part of trunk is thick. It seems that the ancient age and time have made it gradually hollow and divide into two parts, which form the natural tree phase of dual trunks. The opposite gestures between both trunks have brought the vicissitudes of age and historical imprint to the artwork and increased a deep meaning of time and space for the artwork. The artist picks the lower position to bring out the branch and spread to the left main branch according to the characteristics of stump material to strengthen the posture of precipitous wonder and beauty of the space; stabilizing the gravity by the thick branch of the vice trunk spreading to the right brings the artwork back to balance and stable from the steepness. The lines of branches go ups and downs and show the power. The overall posture of branch is spreading or elegant, or flat or uptrending, full of the beauty of change. The random and variable branch lines make the structure full of conflicts and fights of ups and downs. And the artist skillfully drives the trend of the tree vigor and main line and makes the artwork reach the coordination and integration in general, and thus increases the works appreciation content and deepens the connotation.

The out appearance of work is with the grand vigor and elegant gesture. New leaves sprouting all over the tree make people feel like spring breeze which greens the sight. What a warm and sunny scene. The green and graceful inner gesture appears with the indomitable character of pride, inside which are the vicissitudinous experiences and fights of nature and the expression of human's self spirit.

"双雄竞秀" 九里香 *Murraya exotica* 陈伟藏品
——2013中国鼎国家大展首席大奖

"Heroic Duo Contend for Beauty". Jasminorange. Collector: Chen Wei —2013 China Ding Chief Prize of National Exhibition

文：罗传忠 Author: Luo Chuanzhong

陈伟先生的九里香“双雄竞秀”之所以在众多顶级盆景中脱颖而出，高居榜首，荣膺中国盆景国家大展中国鼎首席大奖，自有其过人之处。

其实，盆景界对“双雄竞秀”并不陌生，该作品曾在国家级盆景展览中摘金夺银，并于2006年在陈村国际盆景博览会上作为展会的形象宣传图片而风靡一时。然而，却鲜有人知其成功背后所历经的艰辛和漫长的等待。

据笔者了解，该作品最早下山桩栽培造型是20世纪70年代初，至今已足有40个春秋。作品先是地植，其间曾数次上盆造型及下地复壮。经过20年的蓄枝截干才基本成型，再经10余年的蓄养方才达到目前的水平。此作品作为一个典型的范例，正应了“一景方成已十秋”的行话，其实岂止十秋！如果没有作者日复

点评 Comments ▼

一日、年复一年的精心养护栽培，从青丝熬到白发，何来如此赏心悦目的盆景逸品呢？

不难看出，作品的原桩是最普通的双干树，且主副干左右分道扬镳，是盆景人最忌讳的树相。然而作者似乎早就立定主意，胸有成竹，采取与正格树完全不同的逆向思维开始蓄枝造型。任由左边的主干和右边的副干如脱缰之马，按相反的方向恣意狂奔。副干宛然下行，主干迂回上跃，主副干越走越远，一般人看不出其中奥秘，认为主副干大分大离如何收得拢？殊不知主干以小于90° 的突然急转弯，向中轴线方向急起盘旋而上，结顶向右俯看，照应副干，顾盼有情，礼让得体。此时，才真正显露其庐山真面目，一盆超凡脱俗的盆景终于亮丽地展现在世人的面前。

突破固有的审美标准，不拘常法，不落时趋，创作出令人耳目一新的作品，应该是作者创作该作品的初衷吧。

观此作品，树品高尚，矮壮雄强，气韵生动，聚散有序，节奏和谐，具有强烈的感染力，洋溢着雅趣，玩味无穷；根盘左实右虚，阳根裸露支撑八面，有力稳固；树身洁净，筋骨、肌肉纹理清晰可见，体魄强健，曲线玲珑，身段优美；小枝游弋自如，穿插有序，柔中带刚，过渡自然；叶片细小嫩绿，如繁星点缀，团团簇簇，聚聚散散，碧色满树，美轮美奂；布局有序，疏密得体，层次分明，线条流畅，轻重缓急搭配合适，左跌右飘，左边下跌枝恰好弥补此处的空缺，起稳定重心作用，可谓点睛之笔，使盆树既丰满浑厚又飘逸动感，整体充满奇逸玄妙之气。

Heroic Duo Contend for Beauty of *Murraya exotica* of Mr. Chen Wei has stood out from many top Penjings, ranked the first and won China Ding Grand Prize of National Exhibition of Chinese Penjing because of its special charms.

Actually, Heroic Duo Contend for Beauty is not strange to the Penjing world. This artwork has won many prizes in national Penjing exhibitions, and been popular for a time in 2006 when being on the image promotion photo of Chencun Penjing International Exposition. However, seldom people know the hard and long wait behind the success.

According to the author's understanding, this artwork was firstly cultivated and modeled in the beginning of 1970s, and 40 years have passed. The work was first planted in the earth, and then modeled in the pot and re-cultivated in the earth for several times. The preserved branches and cut trunk has basically formed the shape after 20 years, and reached the current situation after another 10 years of cultivation and cure. As a typical example, this artwork has verified the jargon of "A beauty shall be completed with ten years passing-by." But actually, it always takes more than a decade! Without the elaborate protection and cultivation day after day, year after year, without the hard work which turns the hair from black to silver, how will the elegant Penjing of such beauty be realized?

It is not difficult to see that the original stump of the artwork is the most ordinary dual trunk tree, with the main and secondary trunks separating respectively to the left and right. It is the most unfavorable tree phase for Penjing lovers. However, the artist seems to be determined at the first and has a ready-formed plan in mind to adopt a reverse thinking which is opposite to the positive tree phase and begins to reserve the branches for modeling. He released the left trunk and the right secondary trunk to run widely towards the opposite directions. The secondary trunk creeps down and the trunk wriggles upright. The enlarged distance between the two trunks is further and further... People cannot understand the mystery and wonder how will both trunks be contracted with such big space? They just don't know the trunk suddenly turns by an angle of less than 90°, swirls right up toward the central axis direction, merges at the top and bows down to the right to correlate with the secondary trunk, appearing with the romantic charm and modest manner. Since then, the true spirit of the artwork has been finally revealed, and an otherworldly Penjing has been shown to people at last.

It shall be the artist's original intention when creating this artwork to break through the former standard of appreciation, jump out of the regular rules, catch with the trend and create a refreshing artwork.

When appreciating this artwork, we can see a noble tree character, short but strong with rhythmic vitality. The branches being gathered and released in order with harmonious rhythm are of strong appeal, full of elegant taste and worthy of ruminating. The root base is real at the left and vague at the fight. The positive root is exposed and supportive in all directions firmly and steadily. The tree body is clean with clear and visible bones and muscles. The body is powerful but in exquisite and beautiful shape. The branches stretch freely and cross over each other in order. We can feel the firm in the gentle and the natural transition between them. The leaves are thin and green as the stars gathered and scattered all over the tree forming the splendid beauty. The layout is well arranged with appropriate density and distinguished layers. The lines are smooth and fluent with proper assortment. It falls in the left and rises at the right. The left branch bowing down, which exactly offsets the vacancy and stables the gravity, is the focal to makes the potted tree full and thick while elegant and vigorous. It feels like the whole tree is filled with an aura of wonder and mystery.

中国鼎 China Ding

向下一个时代进军的号角

Horn for Marching to the Next Era

中央电视台第四套国际频道对此次"中国鼎"超级赛事向全球进行了最新中国国家文化新闻的报道

中国鼎——2013(古镇)中国盆景国家大展全景报道

China Ding - 2013 (Guzhen) China National Penjing Exhibition All Report

报道: CP Reporter: CP

中国盆景的下一个时代将走向哪里?

中国盆景近年来的飞速发展已经开始向历史提出了一个重大问题，那就是中国的展览对中国盆景今后的发展走向应该起到什么作用?展览和赛事的价值真的只是让人多拿几个奖牌、大家高兴一下，然后互相展示交流作品的"爱好者之间的活动"而已么?

一个展览该不该有使命感?它对文化该有什么意义?对国家文明该有什么意义?对历史该有什么意义?

"中国鼎"这个词的出现，意义何在?中国已经有了那么多展览，为什么中国盆景艺术家协会(CPAA)还要办这样一个展览品牌?

作为盆景的发源国，中国人应该对自己的盆景文化做一些什么样的具有传承性的事情?这个动作是什么样的?

中国盆景的精神内核到底是什么?这种内核看得到的样子到底是什么样的?如何把这种饱含了民族精神的内核发扬光大，使之成为今后世界盆景文化中有价值的代表中国国家文明的文化遗产?

中国盆景文化应该通过一种什么样的符号化的品牌世世代代地延续下去?100年后让我们的子孙后代说，上100年的中国盆景前辈们没有浪费时间。

同样是做事，殚精竭虑和浑浑噩噩都是做事的方式。中国盆景艺术家协会(CPAA)的第五届理事会团队成员们显然不想当后者。

展场一角

展场一角　展场一角

What will the next era of China Penjing Bring?

The rapid development of China Penjing for the past few years has put forward an important question to the history, which is that, how Chinese exhibitions can affect the future development of China Penjing? Are the true values of exhibitions and competitions to bring joys for everyone just by presenting more awards or of "an event of enthusiasts" just to show their artworks and communicate?

Should an exhibition be of a sense of mission? What is the meaning of it to the culture, to the national civilization and to the history?

What is the meaning of the appearance of word "China Ding"? China has already had so many exhibitions, why does China Penjing Artists Association (CPAA) still hold such an exhibition brand?

As the original country of Penjing, what shall China people inherit for their Penjing culture? How will these gestures be?

What is the exact spirit core of China Penjing? How does this core appear? How to carry forward and further develop this core, which is full of national spirit, and make it become the valuable cultural heritage representing China's national culture in the world's Penjing culture of future?

Through what kind of symbolic brand should China Penjing culture be passed down generation by generation? In a hundred years, our descendants shall say that the predecessors were not wasting their time in the last a hundred years.

Both meditating deeply and muddling along are attitudes for artworks, and apparently the fifth council team members of China Penjing Artists Association (CPAA) don't want to be the latter.

中国盆景艺术家协会(CPAA)第五届理事会团队组织的前所未有的国家级赛事——中国鼎 2013(古镇)中国盆景国家大展，也显然是在通过“中国鼎”赛事向回答以上问题的方向努力。

从一开始出现在展览计划书中的展览品牌的定位，到全国各地的评比甄选作品流程设计，评比规则的一再讨论和考量，展览设计主题的精心思考，展台的配饰、尺寸、色彩等设计的组合方式，展览品牌会徽的主题美术元素……甚至最后的颁奖晚会的内容和表现形式的创意化追求等各个流程阶段，都明显可以看到：中国盆景艺术家协会(CPAA)正在试图对以上这一问题寻找答案。从这个过程中不难发现：“创意”两个字是中国盆景艺术家协会(CPAA)第五届理事会团队成员们的行动关键词，也是解读中国盆景艺术家协会(CPAA)近年来和今后众多重大活动的思维线。

中国盆景艺术家协会会长苏放在开幕式上致辞

“Creativity” is the key word for the activities of the fifth council team members of China Penjing Artists Association (CPAA).

2013 年 9 月 29 日至 10 月 3 日由中国盆景艺术家协会和中山市人民政府联合主办，由古镇镇人民政府、中山市农业局、中山市林业局、中山市海洋与渔业局和广东省盆景协会承办，由中山市南方绿博园有限公司、中山市盆景协会、中山市古镇镇盆景协会协办的中国鼎——2013(古镇)中国盆景国家大展在中山市古镇镇举办并取得圆满成功。

9 月 29 日下午 3:00，开幕式在灯都古镇会议展览中心多功能厅举行。开幕式简约但是不简单，隆重却又不奢华，在贯彻勤俭办展的同时，主办方用中国鼎品牌理念打造中国盆景的划时代明星展会，来自全球的盆景爱好者及国内外近百家媒体网络记者用盆景热情共谱盛况。

世界盆景友好联盟主席胡运骅在开幕式上致辞

中山市古镇镇镇长魏宏锐在开幕式上致辞

中山市人民政府副市长杨文龙在开幕式上致辞

中山市人民政府市委组织部部长雷彪宣布开幕式正式开始

中国盆景艺术家协会会长、世界盆景石文化协会名誉会长苏放，世界盆景友好联盟主席胡运骅，中山市古镇镇党委副书记、镇长魏宏锐，中山市人民政府副市长杨文龙依次在开幕式上致辞。中山市人民政府市委组织部部长雷彪宣布开幕式正式开始。

出席开幕式的领导及嘉宾还有：广东省纪委、监察厅派驻省农业厅纪检组组长、监察员、省农业厅党组成员王力伟，湖北省省纪委监察厅厅长曹志振，中山市人民政府市人大常委会副主任司徒伟湛，中山市人民政府政协副主席马志刚，中山市农业局局长李小建，中山市古镇镇党委书记、人大主席余锡盆，中山市古镇镇党委副书记、人大副主席苏玉山，中山市古镇镇党委委员、副镇长郑海声，中山市古镇镇党委委员、副镇长袁松华，中山市古镇镇人民政府副镇长何新煌，中国盆景艺术家协会名誉会长梁悦美、鲍世骐、马建中，中国盆景艺术家协会名誉会长、世界养生科学联合会会长、国际工商业集团协会主席团主席吴舜龄，中国盆景艺术家协会常务副会长李正银、柯成昆、曾安昌、陈明兴、杨贵生、吴成发，中华盆栽艺术台湾总会、亚太盆栽赏石大会名誉主席陈苍兴，世界盆栽友好联盟中国区主席辛长宝，中国风景园林学会花卉盆景赏石分会顾问李克文，江门市个体私营劳动者协会副会长谢英珠等国内嘉宾，以及近百位来自日本、美国、韩国、捷克、意大利、瑞典、立陶宛、匈牙利、泰国、越南、马来西亚等国的国际盆景著名人士。

世界盆景友好联盟前会长、国际盆栽俱乐部前会长索里塔·罗塞德也在现场认真观摩

展场一角

微型盆景永远是人们的挚爱之一

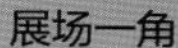
展场一角

展场一角

The unprecedented national event – (China Ding) China National Penjing Exhibition organized by the fifth council team members of China Penjing Artists Association (CPAA) is also an obvious answer to the above mentioned questions through the event "China Ding".

From the first coming exhibition brand positioning in the exhibition proposal to the process design of artworks appraisal and selection all over the country, the repeated discussions and considerations of competition rules, the meticulous thinking of exhibition theme design, the design combination pattern of decorations, sizes and colors, etc., the theme art elements of the exhibition brand emblem ... even to each process stage of seeking for creative contents and expressions form of the award party, we can see that: China Penjing Artists Association (CPAA) is trying to seek the answer to the question mentioned above. In this process, it is easy to see that "Creativity" is the key word for the activities of the fifth council team members of China Penjing Artists Association (CPAA), and also the thinking line to understand the major activities held by the China Penjing Artists Association (CPAA) in recent years and the future.

From the 29th of September to October 3rd in 2013, China Ding-2013 (Guzhen) China National Penjing Exhibition jointly sponsored by China Penjing Artists Association and Zhongshan Municipal People's Government, undertaken by Zhongshan Guzhen People's Government, Agriculture Bureau of Zhongshan, Forestry Bureau of Zhongshan, Ocean and Fisheries Bureau of Zhongshan City and Guangdong Penjing Association, and co-sponsored by Zhongshan Nanfang Green Exposition Garden Co., Ltd., Zhongshan Penjing Association and Zhongshan Guzhen Penjing Association has been held in Guzhen Town, Zhongshan City and achieved a complete success.

At 3:00 p.m. of September 29, the opening ceremony was held in the multi-function hall of Guzhen (City of Light) Convention and Exhibition Centre, Guzhen Town, Zhongshan City, Guangdong. The ceremony was brief but not simple, grand but not extravagant. While diligently and thriftily organizing the exhibition, the sponsor has built an epoch-making star exhibition of China Penjing by the brand concept of China Ding, the Penjing enthusiasts all over the world and nearly a hundred media network reporters have composed the spectacular event with Penjing passion.

Su Fang, the President of China Penjing Artists Association and Honorary President of World Bonsai Stone Culture Association, Hu Yunhua, the President of World Bonsai Friendship Federation, Wei Hongrui, the Deputy Secretary of Party Committee and Town Mayor of Guzhen Town in Zhongshan, and Yang Wenlong, the Deputy Mayor of Zhongshan Municipal People's Government have made speeches at the opening ceremony in turn. Lei Biao, the Minister of the Municipal Organization Department of Zhongshan Municipal People's Government has announced that the opening ceremony officially begins.

The leaders and distinguished guests who have attended the opening ceremony include the domestic guests of: Wang Liwei, the Head of Discipline Inspection Group, Supervisor and Party Leadership Group Member of Department of Provincial Agriculture who is accredited by Discipline Inspection and Department of Supervision of Guangdong Province, Cao Zhizhen, the Director of Hubei Provincial Commission for Discipline Inspection and Department of Supervision, Situ Weizhan, the Deputy Director of the Standing Committee of Zhongshan Municipal People's Congress and Zhongshan Municipal Government, Ma Zhigang, the Deputy Chairman of the People's Political Consultative Conference of Zhongshan Municipal People's Government, Li Xiaojian, the Chief of Zhongshan Municipal Agriculture Bureau, Yu Xipen, the Secretary of Party Committee and Chairman of People's Congress, Zhongshan City, Su Yushan, the Deputy Secretary of Party Committee and Vice Chairman of People's Congress of Guzhen Town, Zhongshan City, Zheng Haisheng, the Party Committee Member and Deputy Town Mayor of Guzhen Town,

Zhongshan City, Yuan Songhua, the Party Committee Member and Deputy Town Mayor of Guzhen Town, Zhongshan City, He Xinhuang, the Deputy Town Mayor of Guzhen People's Government of Zhongshan City, Amy Liang, Bao Shiqi and Ma Jianzhong, the honorary Presidents of China Penjing Artists Association, Wu Shunling, the honorary President of China Penjing Artists Association, President of World Health Science Federation and President of Presidium International Industry and Commerce Group Association, Li Zhengyin, Ke Chengkun, Zeng Anchang, Chen Mingxing, Yang Guisheng and Ng Shingfat, the Executive Deputy Presidents of China Penjing Artists Association, Chen Cangxing, the honorary President of Chinese Taiwan Bonsai Art Association and Asia Pacific Bonsai and Suiseki Convention, Xin Changbao, the President of World Bonsai Friendship Federation in China Region, Li Kewen, the Consultant of Flower Penjing and Suiseki Branch of Chinese Society of Landscape Architecture, Xie Yingzhu, the Deputy President of Private-Owned Workers Association of Jiangmen City, as well as nearly a hundred international Penjing celebrities from Japan, America, Korea, Czech, Italy, Sweden, Lithuania, Hungary, Thailand, Vietnam and Malaysia, etc.

开幕式上嘉宾合影

开幕式现场

开幕式现场

湖北省省纪委监察厅厅长曹志振（右二）和中国盆景艺术家协会名誉会长梁悦美（右一）在开幕式上

And a great many visitors from home and abroad have admired the artworks as "shocking", "satisfying", "eye opening", " fascinating" and "beyond expectation"...

本次展览的展场是中国盆景展览历史上最大的一次室内展展场——灯都古镇会议展览中心，总面积超10000m^2。展场中的宏大气派国内罕见，展场布置的纹饰设计和展台具有中国传统文化风格的桌旗首次出现，细节中的新元素也令人耳目一新。

而出现在中国鼎中国盆景国家大展上的100盆中国顶级盆景形成了中国盆景最新的国家级盆景展的入列水准的亮丽风景线！这100盆中国顶级盆景（见专辑2013（古镇）中国盆景国家大展）都是首次以"国家大展入选人"的骄傲身姿站立到了2013年中国盆景的明星行列中。其中的许多展品均是首次出现在国家级展览的站台上。令很多国内外参观者大呼"震撼"、"过瘾"、"开眼了"、"太迷人了"、"没想到"……

来自陈伟先生的"双雄竞秀"获得了万众瞩目的中国盆景国家大展的首次超级大奖——中国鼎。另外8盆特别突出的盆景作品则获得了仅次于中国鼎大奖的中国盆景国家大展奖。有的展品总评分距离中国鼎甚至只差0.2分，一步之遥！

The hall for this exhibition is the biggest indoor exhibition hall in the Penjing Exhibition history of Guzhen (City of Light) Convention and Exhibition Centre, with an area exceeding 10,000m^2. The grand style of exhibition is rarely seen in China. The new elements in details such as the decorative design of the hall layout and the first appearance of table flag on the exhibition stand which is full of China traditional culture style are refreshing.

And the 100 China top Penjing in the China Ding China National Penjing Exhibition has formed the beautiful scenery

开幕式现场

2013 中国古盆收藏展展区

2013 中国盆景艺术家协会会员赏石展展区

中国盆景艺术家协会会员展一角

中国盆景国家大展展区一角

which also represents the listed level of the latest China National Penjing Exhibition! It is the first time for the 100 China top Penjing (the issue of 2013(Guzhen)China National Penjing Exhibition) to appear as stars of 2013 China Penjing with the title of "National Penjing Exhibition Selection". It is also the first time for many exhibits to be shown on the stage of national exhibition. And a great many visitors from home and abroad have admired the artworks as "shocking", "satisfying", "eye opening", " fascinating" and "beyond expectation"....

"Heroic Duo Contend for Beauty" of Mr. Chen Wei has won the much-anticipated first grand prize of China National Penjing Exhibition – the China Ding. Other extraordinarily outstanding 8 Penjing have won the Prize of China National Penjing Exhibition which is second only to the grand prize of China Ding. The total score of some exhibit is only 0.2 point less than China Ding, only one step beyond!

中国盆景艺术家协会会长苏放在展场上接受采访

捷克《盆栽》杂志主编斯瓦托普卢克·马特杰卡对岭南派技法的盆景兴趣浓厚

除主题展会中国盆景国家大展之外，还同步举办了2013中国盆景艺术家协会会员精品展，200盆会员展作品中有的展品的水平甚至不亚于任何国家级展览的金奖展品，此次会员展也评出了2013中国盆景会员展年度大奖1名、2013中国盆景会员展金奖10名、2013中国盆景会员展银奖20名、2013中国盆景会员展铜奖40名（见本书获奖作品专栏）。

此次大会同期的2013中国古盆收藏展展出了46件古盆展品（见本书2013中国盆景会员展之古盆欣赏），许多博物馆级的历史性古盆也是在中国盆景展中首次出现。由于中盆协的许多会员也是赏石爱好者，所以与盆景同时，2013中国盆景艺术家协会会员赏石展还在此次大会上展出了52件世界一流的赏石作品（见本书2013中国盆景会员展之赏石欣赏）。

为了大力推动盆景市场的发展，此次大会还推出了56个销售展位的2013中国盆景秋季贸易周活动。

展会期间，还举办了中国盆景艺术大师樊顺利、徐昊的现场制作表演及中国盆景艺术家协会高级技师陈万均的现场盆景互动教学。

中国盆景年度之夜颁奖晚会在29日晚使此次大会掀起了新的高潮，众多令人耳目一新的颁奖内容和节目让现场的众多国内外嘉宾大开眼界（见2013中国盆景年度之夜特别专辑）。

展览之后，中国盆景艺术家协会还组织了国内外嘉宾团参观访问了深圳的“趣怡园”、东莞的“真趣园”、西安的“唐苑”三个中国的私家盆景园林（见本书2013中国鼎之旅）。

EBA欧洲盆栽协会波兰分会会长沃齐米日·皮特思科也对岭南派技法的盆景兴趣浓厚

与历年来国内外盆景展览不同的是：1.本次展览设计采用代表了中国传统文化和文物收藏符号的青花瓷元素作为大会和展台的设计主题，很好地结合了中国古典文化元素与现代设计理念；2.中国盆景国家大展展区为独立展台单独展示，这种为每一盆展品打造专属空间的做法是世界展中首创，向全世界传达了盆景已经进入高端艺术品陈列展台的中国最新信息；3.评委信息和评比结果公开并公示，在逐步透明化的过程中探讨今后中国盆景评比的公平公正的新方向；4.减少奖项、提高奖金，从而提升获奖作品的含金量，本次获奖数量大幅度减少，许多人虽然没有拿到奖但也很高兴地说：我的盆景能入选本身已经说明很多问题了。

世界盆景友好联盟前财务长、美国国家盆景基金会前副会长蔡斯·罗塞德在展场上

评委认真开展评比工作

樊顺利大师现场制作表演

徐昊大师现场制作表演

日本景道家元二世须藤雨伯与中国盆景艺术大师陆志伟在展场上交流

2013 Chinese Ancient Pot Collection Exhibition held simultaneously as the event has exhibited 46 ancient pots.

Besides the theme China National Penjing Exhibition, 2013 China Penjing Member Exhibition of China Penjing Artists Association has been held simultaneously.Same of the 200 of members' artworks were as fine as the golden prize exhibit of any national exhibition In Ching Penjing Member Exhibition Of China Penjing Artists Association I Annual Grand Prize, 10 Golden Prizes, 20 Siluer Prizes and 40 Branze Prizes of China Penjing Member Exhibition have been presented (see the special cdumn for award-winning artuorks in this book).

2013 Chinese Ancient Pot Collection Exhibition held simultaneously as the event has exhibited 46 ancient pots (see the Ancient Pot Appreciation of 2013 China Penjing Members' Artworks Exhibition in this book). Many historic ancient pots of the museum quality have been exhibited for the first time in China Penjing Exhibition. Since many members of CPAA are stone appreciation lovers, in the Members' Stone Appreciation Exhibition of China Penjing Artists Association 52 world's top stone appreciation artworks have been exhibited with the Penjing (see the Stone Appreciation of 2013 China Penjing Member Exhibition in this book).

In order to vigorously promote the development of Penjing market, this exhibition also has sponsored the activity of 2013 China Penjing Fall Trade Week with 56 sales booths.

During the exhibition, the live demonstration performed by Fan Shunli and Xu Hao, the great artists of China Penjing and the live Penjing interactive teaching performed by Chen Wanjun, the senior technician of China Penjing Artists Association, have also been held.

The Annual Night Award Party of China Penjing in the evening of 29 has set up a new upsurge for this event. The numerous refreshing prize presenting content and performances have greatly broadened the horizons of all the guests at home and abroad (the special issue of 2013 Annual Night of China Penjing).

After the exhibition, China Penjing Artists Association has also organized the domestic and foreign guests group to visit three Chinese private Penjing gardens as Quyi Garden in Shenzhen, Zhenqu Garden in Dongguan, Tangyuan Garden in Xi'an (2013 China Ding Trip in this book).

The event is different from the domestic and foreign Penjing exhibitions over the years by that I. It has adopted the blue and white porcelain element, which represents the Chinese traditional culture and cultural relics' collection symbol, as the design theme of the event and exhibition stand in the exhibition design, which has well combined the China classical culture elements and the modern design concept; II. The exhibition area of National Exhibition of China Penjing is an independent exhibition stand for sole exhibition, and it is the first time in the world to build an exclusive space for each exhibit and has sent the news to the world that China Penjing has entered the high-end art display exhibition; III. The judges' information and comparison results are open and publicized, and the new direction of justice and equity in China Penjing rating has been discussed during the gradually transparentized process; IV. Reduce the awards and raise the bonus to enhance the gold content of the awarded artworks. The awarded number in this exhibition is greatly reduced, and many people who haven't won the prize happily said that My Penjing being selected has explained a lot.

中国盆景艺术家协会常务副会长杨贵生与外宾们在展场上交流

2013 中国盆景年度之夜

中国盆景国家大展的首次超级大奖——中国鼎
“双雄竞秀”九里香 陈伟藏品

2013 中国盆景国家大展奖“雄风”
刺柏 鲍世骐藏品

2013 中国盆景国家大展奖
“奇劲唱风” 赤松 曹志振藏品

据统计，本届为期 6 天的展期总计吸引了近 8 万人次参观，共有 162 位来自各个省份的盆景送展人参展，展出的展品总价值超 3 亿人民币，展览期间所展示的盆景交易金额达到 3000 万元以上，盛况空前！年度颁奖晚会这一盛典更是群星闪耀。晚会上上共颁布 10 余类奖项，包括本届展览的中国盆景国家大展奖、中国盆景艺术家协会会员展年度大奖、中国盆景艺术家协会会员展金奖、2013 年度个人单项奖、2013 中国盆景年度城镇、2013 中国盆景年度协会等，其中“中国鼎”首席大奖获得者荣获价值 30 万元的玉质奖杯一届保存权、8 万元人民币奖金及“2013 中国盆景年度先生”荣誉称号。

2013 中国盆景国家大展奖
“疑是枝头蝶恋花” 簕杜鹃
吴成发藏品

2013 中国盆景国家大展奖
“龙飞凤舞” 榕树 李正银藏品

中国鼎国家大展的成功举办，显然只是中国盆景艺术家协会推出的“国家大展”品牌序列的开始，序列中的第二个子品牌：中国尊——中国盆景收藏家国家大展将成为 2014 年中国国家级盆景展中的下一个焦点。届时，众多中国著名的盆景收藏家将参加这一轮以个人收藏品群体质量为评比对象的新一轮有创意的国家级大赛，为此，在 29 日的“中国盆景年度之夜”晚会上，2014 年首届中国尊——中国盆景收藏家国家大展的举办地也成了当晚晚会的焦点，浙江省余姚市市政府和浙江余姚的中国盆景艺术家协会名誉会长马建中先生的高风中学成功申办“中国尊——2014（余姚）中国盆景收藏家国家大展”。中国盆景艺术家协会会长苏放在晚宴上，将中国盆景艺术家协会的展览会旗从中山古镇转交到了马建中先生手中（见专辑三年度之夜专辑）。

CCTV13 中央电视台新闻频道、CCTV4 中央电视台中文国际频道、央视网、《南方日报》、《中山日报》、中山电视台、中山电台、古镇政府网、《台湾盆景世界》网站、盆景乐园网、岭南盆景网等上百家国内外媒体对本次展会进行报道，日本《近代盆栽》、美国《国际盆栽》、法国《气韵盆栽》、意大利《盆栽与新闻》、西班牙《当代盆栽》、捷克《盆栽》杂志等近几十家国外专业盆景媒体网站也将陆续刊出本次展览专题报道。

此次中国鼎国家大展的完美展示让全球盆景人为之注目，为世界盆景舞台贡献了一个全新的来自中国的国家盆景展品牌，它的成功举办向全世界发出了一个信号，那就是“中国的新盆景时代”已经开始。

2013 中国盆景国家大展奖
九里香 罗崇辉藏品

2013 中国盆景国家大展奖
真柏 陈国健藏品

2013 中国盆景国家大展奖
“盛世” 五针松 杨贵生藏品

2013 中国盆景国家大展奖
“天娇” 真柏 陈伟藏品

中国盆景艺术家协会会长苏放（左）和古镇镇长魏宏锐（右）为中国鼎国家大展首席大奖获得者陈伟（林文镇代领）颁奖

中国盆景艺术家协会会长苏放（右）和中国盆景艺术家协会名誉会长马建中（左）进行会旗交接

嘉宾参观趣怡园和园主吴成发（右）合影

嘉宾参观真趣园和园主黎德坚（右二）合影

According to the statistics, the exhibition period of 6 days has attracted nearly 80,000 person-times to visit, there were totally 162 Penjing exhibitors from each province, the total value of exhibits exceeded 300 million yuan, the trade amount of the exhibited Penjing during the exhibition period reached more than 30 million yuan, it was an unprecedentedly grand occasion! The grand ceremony of the annual award-presenting party was star-studded and shining all over China. More than 10 awards have been presented at the party, including the Prize of China National Penjing Exhibition, the Annual Grade Prize of China Penjing Member Exhibition of China Penjing Artist Association, the Golden Prize of China Penjing Member Exhibition of China Penjing Artist Association, the 2013 Annual Individual Prize, the 2013 Annual Town of China Penjing and the 2013 Annual Association of China Penjing, etc., in which, the winner of the grand prize "China Ding" has the honor to win the retention right of the jade trophy of 300,000 yuan, bonus of 80,000 yuan and the honorary title of the "2013 Annual Mister of China Penjing".

嘉宾团参观

嘉宾团参观

The success of China Ding National Exhibition is apparently the beginning of the "National Exhibition" brand serials promoted by China Penjing Artist Association. The second sub-brand of the serials:

China Zun –China Penjing Collectors National Penjing Exhibition will be the next focus in 2014 China National Penjing Exhibition. Numerous Penjing collectors of the most famous in China will participate in the new round inventive national competition in which the individual collection group quality will be the appraisal objects. Thus in the night party of "Annual Night of China Penjing" on 29, the host place for China Zun - China Penjing Collectors National Penjing Exhibition has become the focus of the night. Yuyao Municipal Government of Zhejiang Province and Gao Feng High School founded by Mr. Ma Jianzhong, the honorary President of China Penjing Artists Association from Yuyao, Zhejiang has successfully bidden "China Zun – 2014 (Yuyao) China Penjing Collectors National Penjing Exhibition". Su Fang, the President of China Penjing Artists Association has handed the exhibition flag of China Penjing Artists Association from Guzhen, Zhongshan to Mr. Ma Jianzhong (the special issue of 2013 Annual Night of China Penjing).

More than a hundred domestic and foreign medias such as CCTV News Channel at CCTV 13, CCTV International's China Channel at CCTV 4, CNTV, Nanfang Daily, Zhongshan Daily, Zhongshan TV Station, Zhongshan Radio Station, Guzhen People's Government Website, Taiwan Bonsai World Website, Penjing Paradise Website and Lingnan Penjing Website, etc. have reported this exhibition. And dozens of foreign professional Penjing media websites, such as *KINBON* in Japan, *International Bonsai* in America, *Esprit Bonsai* in France, *Bonsai and News* in Italy, *Bonsai ACTUAL* in Spain and *Bonsai* in Czech, etc. will publish the special reports on this exhibition successively

The perfect performance of China Ding National Exhibition has attracted the attentions of Penjing lovers all over the world and contributed a brand new national Penjing exhibition brand from China to the world Penjing stage. Its success has sent a signal to the whole world that the "New Era of China Penjing" has begun.

"雄风" 刺柏 *Juniperus formosana* 鲍世骐藏品 ——2013中国盆景国家大展奖

"Heroic Spirit". Taiwan Juniper. Collector: Bao Shiqi — 2013 China National Penjing Exhibition Award

文：王选民 Author: Wang Xuanmin

在本届大展中刺柏"雄风"是一件光彩夺目的作品。它的艺术表现力具有强烈的个性风彩。可谓是标奇领逸的言鼎之作！其作品的成功创作有以下特点；

1. 首先，作者在素材的选择上具有独特的眼光。早期看这棵树仅是普通的刺柏桩材而已，从根盘上至主干延续分枝并无明显审美优势。平庸化神工，而今是造型完美，从局部到整体处处都有审美亮点，真是难能可贵非一般作者能为。

2. 作品审材取势，把素材特点发挥到极致。在确立树相时胸中自有千年古柏的健骨之气，苍润之美。故得今日"雄风"之大气风范。

3. 作品应用雕刻，整型之特殊造型技法，做到了技术为艺术效果服务，意匠参造物创造了"雄风"之大观。

4. 作品章法布局，安排得当。在枝叶之间营造了层次，利用层次的组合营造了空间，在空间穿插了气息，在气韵流转中动势得到了延伸，故"自然之神"油然而生。

5. 作品将刺柏的叶性和枝的完美结合从而得到完美的表现。松柏类盆景造型对于叶性美和枝性认识要求非常高，并非每一件作品都能得到正确和良好的表现，这在品评每一件作品时至关重要。

6. 作品的制作培养功力已达十年之多，造枝的成熟度已进入最佳观赏期。

In the exhibition, the China Savin "Heroic Spirit" is a piece of glamorous artwork. The artistic expression is of a strong personality style. It can be appraised as an unconventional and elegant artwork of the top level! The successful creation of this work has the following characteristics below.

Firstly, the artist has a unique opinion on the selection of material. This tree is just an ordinary China Savin stump at the first sight. There is no obvious beauty appreciation advantage from the root base up to the trunk till the branches. At last, the mediocrity turns into the God's artwork with the perfect modeling, the appreciation highlights can be found everywhere from the parts to the whole. The ordinary people can't accomplish such an artwork.

The material of the work is depending on the gesture, and the material characteristics have been developed to the most. When determining the tree style, the artist owns the grand spirit and vigorous and full beauty of the ancient China Savin of a thousand years old in his heart, and so forms the grand style of "Heroic Spirit".

This work is formed by the application of special modeling techniques of carving and shaping. The artist has utilized the technology to serve the artistic effect and created the grand view of "Heroic Spirit" by redesigning the reference object.

The layout of the work is well arranged. Levels have been built between the leaves and branches and the combination of levels creates space. The dynamic gesture gets to extend among the transfer of the artistic conception and so forms the "romantic charm of nature".

The work perfectly expresses the combination of leaves and branches of China Savin. The beauty of leaves and the knowledge of branches are very important to the Penjing modeling of conifers. Not every piece of work can be correctly and well expressed. It is crucial when appreciating every piece of work.

点评 Comments

“雄风”刺柏 *Juniperus formosana* 高102cm 宽120cm 鲍世骐藏品 苏放摄影
"Heroic Spirit". Taiwan Juniper. Height: 102cm, Width: 120cm. Collector: Bao Shiqi, Photographer: Su Fang

“奇劲唱风”赤松 *Pinus densiflora* 曹志振藏品

——2013 中国盆景国家大展奖

“The Wind Blowing with Powerful Vigor”. Japan Red Pine. Collector: Cao Zhizhen

—2013 China National Penjing Exhibition Award

文: 徐昊 Author: Xu Hao

高山之巅，苍穹之下，古松傲立，岚烟明灭。

微风轻轻的吹过，山间的虫儿在奏乐，鸟儿在高歌，山巅的古松也和着微风低吟浅唱，天籁之音穿过岚烟，越过叠嶂，划向悠远的清空。忽然间， 风云变幻，狂风卷起山谷的云雾，如潮水般向山顶涌来,瞬间便吞没古松,淹没群峰。虫儿们缩进缝隙中无声无息，鸟儿也不知躲去了哪里，唯有古松却雄壮地高歌起来，伴随着万松齐鸣，声如大海波涛，势如万马奔腾，震荡山谷，响彻云霄。风卷云过，古松从容不迫，巍然屹立，沐浴在万道霞光中，依旧低吟浅唱。这就是大展奖作品“奇劲唱风”展现给我们的意境和内涵之美。

作品根部由两条粗大腾空的主根组成，主根近盆面处各分侧根数条，稳扎地面，根线具有转折、疏密的变化。树干苍古雄壮，扭筋转骨，充满力量感，主干自中部向右横折，复又转折向上，转折几乎成直角。这种转折角度平常会被盆景创作所忌讳，但反映在如此古雅的树干上，不仅显得自然而然，而且增强了松树作品的“奇劲”之气，同时也深化了作品的精神内涵。作品取势向右，因此在布枝时将主枝作向右的凌空舒展状，主枝的线条以直线转折来增强古松的劲健之气，复以主枝同位的小枝作向上的扭曲转折状，以此衬托主枝，使之产

“奇劲唱风” 赤松 *Pinus densiflora* 高 140cm 宽 130cm 曹志振藏品
"The Wind Blowing with Powerful Vigor".
Japan Red Pine. Height: 140cm, Width: 130cm. Collector: Cao Zhizhen

生错落变化之美。树冠布枝茂密，间有疏密虚实，与雄壮的躯干相协调。左下方以一背枝左折作点缀，枝线与主干相交，枝端向左下方舒展，枝片浑厚，使得原本重心右倾的树姿复归稳定，呈现安然之象。

奇松有奇骨，作品根如巨足，干似腾蛟，看似历经沧桑，仍却奇劲苍茂，真有"自信岁月三千载，风霜雨雪唱不休"之势。

The Wind Blowing with Powerful Vigor

The extraordinary pine has rare bones. The root of work looks like a pair of giant feet, and the trunk a soaring dragon. It seems to have gone through all the vicissitudes but still stands vigorously and green. It just has the imposing manner of standing in confidence for 3,000 years and singing unceasingly in the wind and snow.

On the top of mountains and under the dome, the ancient pine stands and the mist flickers.

When the wind gently breezes, insects buzz as music in the valley and birds sing. The ancient pine stands on the mountain top also hums with the wind. The sound of nature flies through the morning mist, over the peaks and toward the remote sky.

Suddenly, the wind blows hard and changes the form of cloud. The fierce wind rolls the mist in the valley and pushes it towards the mountain top as the wave, which swallows the ancient pine and submerges the peaks. Insects hide into cracks in silence and birds cannot be seen anywhere. Only the ancient pine majestically sings with the pine forest. The sound is like the ocean tides and thousands of horses running by. It shocks the valley and echoes to the clouds. After the wind rolls the clouds over, the ancient pine still stands calm and firm, bathing in thousands of sunshine rays and singing with low voice.

This is the beauty of artistic conception and contain revealed by the exhibition prize winner Vigorously Singing in Wind.

The root part of work is composed by two thick soaring trunks, which are respectively divided into many strips near the pot top firmly rooted in the earth. The root lines are with changes of direction and density. The tree trunk is old and vigorous. The twisted trunks and branches are full of power. The trunk folds from the center to the right transversely and then bends upright again by a nearly right angle.

Bending angle of this kind is usually unfavorable in Penjing creation, but reflecting on such a quaint trunk not only seems natural, it can add the extraordinarily vigorous spirit to the pine work and deepen the spirit containing of the work. The gesture of work is to the right. Thus during branch layout, the main branch is made as spreading to the right. The line of main branch uses the straight line bending to increase the powerful spirit of the ancient pine. Combining with the branches at the same position twisting upright to emphasize the trunk, a beauty of random changes is thereby created.

The branch layout of tree crown is dense with occasional changes which coordinate with the robust trunk. The left lower part is dotted with a back branch bending to the left. The branch line intersects the trunk and the branch end spreads to the left lower part. The thick and rich branch re-stables the tree gesture which is with the gravity leaning to the right, and presents a calm and comfortable view.

The extraordinary pine has rare bones. The root of work looks like a pair of giant feet, and the trunk a soaring dragon. It seems to have gone through all the vicissitudes but still stands vigorously and green. It just has the imposing manner of standing in confidence for 3,000 years and singing unceasingly in the wind and snow.

“疑是枝头蝶恋花” 簕杜鹃 *Bougainvillea spectabilis* 高120cm 吴成发藏品
“Just Like the Circling Butterflies above the Tree Top Staying for the Flowers”. Leafyflower. Height: 120cm. Collector: Ng Shingfat

点评 Comments

"疑是枝头蝶恋花"簕杜鹃 *Bougainvillea spectabilis* 吴成发藏品 — 2013中国盆景国家大展奖

"Just Like the Circling Butterflies above the Tree Top Staying for the Flowers". Leafyflower. Height: 120cm. Collector: Ng Shingfat — 2013 China Ding China National Penjing Exhibition Award

文：李奕祺 Author: Li Yiqi

簕杜鹃（三角梅），树身主干粗壮扭曲，苍劲有力。树皮裂纹处处，还有若干孔洞疤痕，是岁月累积的造化。树上三五横枝，左右错落，由粗变细，如巨臂婉转向下。主干与横枝之间，长短粗细，比例合理，主次分明。那些无数垂落的细枝，虽然密密麻麻，但并不杂乱，而是枝向协调一致，有条有理，垂落有序，且密中有疏，留下不少空白，疏密交错，展现出天然垂杨柳的美态。

经过人为的刻意造型，此簕杜鹃已完全改变了其本身的自然树相，形成垂杨柳的缩影，创造性地发明了树种枝法脱胎换骨变幻的跨越。它形似垂杨柳，但别于垂杨柳，更胜垂杨柳。深秋之后，花朵盛开，满树紫红色，犹如紫霞垂照。清明时节，残花雕落，枝头冒出星芽点点，不久又是绿叶成阴，又开始进入新一轮的营养生长。如此这般"垂杨柳"，堪称一绝，是旷世罕见的珍品。

造就如此珍品，工程浩大。那些不计其数的下垂细枝，每一条都经过精心处理，必须逐条扭转向下，捆绑固定，假以时日，定型之后，方可解套。工作量之大，需夜以继日方能完成。而成型之后的长期养护，为求保持原貌，同样困难。那是因为必须面对密密麻麻的下垂枝，不断修剪向上芽，逐枝扭下并固定好新生枝，蟠虬定型之后，方能功成松绑。稍有松懈怠慢，就会变形走样，引致枝条杂乱，破坏原貌。如此艰辛烦琐，非一般人能长期承受。由此可见，这种珍品之难得可贵。

遥望此珍品，恰似秀发披肩女郎的背影，矗立在湖畔长廊，人们都期盼她回眸一笑。在湖对岸，天边朦胧的山峦，玲珑浮凸，体态优美，隐含着睡美人的神秘和羞涩，与湖中倒影交相辉映，与秀发女郎遥遥相对。

水面上，湖光粼粼，游人小船划过，扩散出圈圈波纹，轻轻划破了谧静。湖边绿色水浮莲，冒出一串串紫花，随波微微浮荡，分外柔和添彩。

长廊中，游人闲庭漫步。有人驻足向湖中投食喂鱼，引来一大群锦鲤浮游水面，翻滚争食，顿时呈现出一团色彩斑斓、活泼灵动的画面，吸引众人观赏。

年轻恋人偎依树下，绵绵细语，情话连连，沉浸交融之情景，陶醉浪漫之情怀。

Leafyflower has thick twisted vigorous and forceful trunk. Bark cracks everywhere and numbers of holes scar show its years. There are several lateral branches, strewn at random, from thick to thin, such as macrobrachia downward. Between the trunk and lateral branches, ratio of branches' size is reasonable; the primary and secondary are clear. The twigs of countless hung, dense but not mixed and disorderly, but even to coordinated, orderly, and sparse in dense with many blank, show the beauty of natural willow.

After artificial deliberately modeling, the Leafy flower has completely changed its own natural phase, forming a microcosm of willow creatively invented the huge change of the species and technique. It is shaped like a willow, but different with the willow, even better than the willow. After the late autumn, the flowers are all in full bloom, full purple, like sunset glow shine down. In Qingming Festival the flower faded, the buds sprouted from the branches then turn to many leaves for shadow soon, after that a new round of vegetative growth is coming. Such willow is extraordinary rare treasures.

It's an enormous project to make such a treasure. Those countless prolapsed twigs, each of them should be manage carefully, turn down and then fixed them using bounding, after some time release them. Such a huge work need continues make day and night. It's hard to long-term maintenance after making for keeping its shape. Keeping dense droopy branches need prune shoots on the top of branches, turn branches down and then fixed them. If not attention, branches disordered and shape deformations make the original destroy. It's so hard and tedious that ordinary people can't afford for a long time. Thus, this is a kind of rare precious treasures.

Look at the treasures, it like the back of hair shawls girl who stands in the corridor near the lake, people are looking forward to her looking back with a smile. On the contrary of lake, the mountains keep close to the sky, with exquisite and graceful posture, which have mysterious and shy of the sleeping beauty, hand in photograph reflect with the reflection in the lake as opposed to a long hair girl.

Tourist row boat across the lake make waves spread out of circles, which break the silent of water surface. String purple flowers emitted from water hyacinth on the lack, which float by the waves slightly, form a beautiful sense.

Visitors walk along the long profile. Someone stopped to hurl food feed the fish in the lake, which attract a large group of fancy carp floating on the water. They dodged that bring a colorful, lively and smart picture which attracts people to watch.

Young lovers snuggle under the tree with the gentle whisper. What an immersed blend of scene makes people intoxicated with the romantic feelings!

“龙飞凤舞” 榕树 *Ficus microcarpa* 李正银藏品——2013中国盆景国家大展奖

“Like Dragons Flying and Phoenixes Dancing”. Smallfruit Fig. Collector : Li Zhengyin — 2013 China National Penjing Exhibition Award

文：李奕祺 Author: Li Yiqi

榕树，整棵树扇形展开，高120cm，宽220cm，左右对称，冠幅颇大。主干粗壮且短，直径达40cm，但高度只有15cm，茎与根紧连，根爪攫土，力度超强，霸气非凡，几近夸张，依此比例，是典型的矮仔大树型。

如此粗大矮壮、雄浑健茂、重心内敛、坐地稳固的大型古树，在自然界，时有所见。若在恶劣环境下，即使面临绝世台风依然可以屹立不倒。可以想象：人立树下，必然感受到巨树压顶，铺天盖地，荫翳华盖，有慑人心魄之震撼。

主干之上，约20条粗大支干，弯弯扭扭，线条浑厚，婉转辐射出去，犹如苍龙飞舞，空群出动，以千钧之力迸发出去，力之所及，所向披靡，它阳刚雄健，气度浩然，流畅奔放。

树梢有无数细枝，分布四处，松而不散，疏密有序，如无数雀爪伸展，柔软纤细，点缀着树冠扇形的弧形边缘，细枝在刚强树干的边缘轮廓上展现妩媚的柔态，造就榕树刚中有柔，柔中有刚，刚柔相生并济的自然景观。

中国岭南地区的榕树，冬天并不落叶，是四季常青的灌木。由于新枝不断繁衍，树体不断扩大，形成一片，独木成林，绿叶成阴。以整体角度看，粗壮超短，主干是主，近20条支干是次，主次分明；深入林中，则每条支干均可被看成独立的一棵树，支干变成主干是主，末梢细枝是次，也是主次分明。由此可见：主与次是相对而言。榕树独木成林时，次干变主干，是别样的主次分明。

榕树常常吸引百鸟来树上栖息，尤其湿地旁的榕树，更吸引群鸟在树上筑巢繁衍。

清晨，霞光初照，白鹭出巢，夜游灰鹭归巢，漫天飞鸟，翱翔飞舞，嘎嘎声交鸣，蔚为壮观；黄昏，夕阳西下，晚霞染红了天边，整天在外觅食的白鹭，倦鸟知返；灰鹭离巢，刚刚开始夜晚的猎食，寂静了一天的榕树上空，顿时又热闹了起来，百鸟盘旋，莺歌燕舞。

小鸟栖息树上，啄食树上害虫，促使树木健康生长。鸟树相依，和谐相处，榕树就是人们心目中的“小鸟天堂”。

鹭鸟朝出晚归，或者暮出晨归。鸟类世界原来同人类社会差不多，生活之中不经常“有人辞官归故里，有人屡夜赶科场”么？

Smallfruit Fig, with the whole tree spreading as the fan-shape, is 120cm heighth and 220cm width. It is bilaterally symmetrical with relatively big crown diameter. The trunk is thick and short with the diameter reaching 40cm, but only 15cm length. The stem is tightly connected with the root, and the root claws grab the earth with powerful strength. The gesture is domineering with exaggeration. It belongs to the typical short and large tree type according to the scale.

A large size steadily rooted ancient tree which is so thick and short with great body weight, powerful and healthy leaves and restrained gravity is occasionally seen in the nature. In case of severe environment, it can stand upright even when encountering the strongest typhoon. We can imagine that when standing under the tree, one may feel the weight of the tremendous tree, and the whole area will be covered by its leaves, the feeling of shock will be breathtaking.

Above the trunk are about 20 thick branches twisted with powerful lines and radiating out like dragons dancing. They burst out with tremendous and invincible force. The tree is masculine and powerful with grand spirit and unrestrained fluency.

There are countless branches on the tree top distributing everywhere, loose but not scattered, and with orderly density. It

点评 Comments

is just like the soft and thin birds' claws spreading out and decorating the fan-shaped arc edge of the tree crown. The branches show in the charming and soft gesture on the edge contour of the strong trunk, and that creates the natural scenery of Smallfruit Fig as firm and gentle combined together.

The leaves of Smallfruit Fig in Lingnan Area of China don't fall in the winter. It is an evergreen shrub because the new branches continuously grow and the tree body enlarges unceasingly and covers a large area. One Smallfruit Fig can form a little forest with green leaves. Appreciating in the whole point of view, it is thick and especially short and distinguished for the primary trunk and the nearly 20 secondary branches deep in the forest, each branch can be seen as an independent tree with the primary and secondary distinct branch and the secondary branches. Therefore, the primary and the secondary are relatively variable. When the Smallfruit Fig forms a forest by only one tree, it is a different kind with the secondary trunk as the trunk.

The Smallfruit Fig usually attracts birds to rest on it. Especially for the Smallfruit Fig near the wetland, it is more attractive for the birds to nest and breed on.

In the morning, when the sunlight shines early, egrets are out of nest and the Grey Heron wandering in the night return to the nests. The view of birds flying and dancing in the sky quacking to each other is magnificent. And in the evening, when the sun goes down and the sunset glow paints the horizon in red, egrets searching food outside fly back home. Then the Grey Heron leave and begin their night hunting. The silent sky above the Smallfruit Fig suddenly becomes lively again with a hundred birds circling around and orioles singing and swallows dart-the joys of spring.

The birdies stay on the tree and peck the insects from the tree to help the tree grow healthily. The birds and the tree live in harmony. The Ficus microcarpa is the "Birds' Paradise" in people's heart.

The egrets either leave early or return late. The birds' world is not very different from ours. Is there a saying "someone resigns to return home, while someone rushes for the officialdom" in the real life, isn't it?

"龙飞凤舞" 榕树 *Ficus microcarpa* 高120cm 宽220cm 李正银藏品
"Like Dragons Flying and Phoenixes Dancing". Smallfruit Fig. Height: 120cm, Width: 220cm. Collector: Li Zhengyin

九里香 *Murraya exotica* 高128cm 罗崇辉藏品
Jasminorange. Height: 128cm. Collector: Luo Chonghui

点评 Comments ▼

九里香 *Murraya exotica* 罗崇辉藏品 —— 2013 中国盆景国家大展奖

Jasminorange. Collector: Luo Chonghui —2013 China National Penjing Exhibition Award

文：张志刚 Author: Zhang Zhigang

盆景创作师法造化，应不拘一格，随性而发。只有“因情而动，因材施艺，见机取势，用心所成”的作品方能独具特色，饱含诗情，才会打动人。罗崇辉先生的九里香作品情景交融、神形兼备，能在首届国家大展中脱颖而出，荣获佳绩，就是一很好例证。

作品主体选用的是一山采古桩，历经百年沧桑，已大面积枯朽，即便残存的三分之一树干也支离破碎，摇摇欲坠。然而作者匠心独运，通过数十载的精心呵护、蓄枝截干，不但使其复壮，而且脱胎换骨，重塑枝冠，再造了一神奇古木风姿，这种化腐为神的手法，凸显了岭南盆景独到技法的艺术魅力。

作品是棵树，自然复古，但又像一有生命的雕塑，怡情传神。左右分合的主干有如两夫妇相依相偎，漫步起舞，虽至耄耋之年，依然绽放生命的浪漫，给人爱的启迪和美的教育。

Penjing creation and imitation shall not be limited to one shape, but be developed along with the nature. The artwork with “moving for affection, application of proper technique, deployment according to circumstances and wholeheartedly making” will be unique, poetic and touching. There is a good example - Mr. Luo Chonghui's “Murraya exotica” artwork, with the fusion of feelings and the natural settings, the combination of both appearance and spirit, it stands out from the first National Show and wins a good result.

The main body of the artwork is an ancient pile from the mountain: After one hundred years, a large area is withered and rotten, and even the residual 1/3 trunk is also fragmented and teeters. However, with the master's ingenuity and care of “branch growing and trunk cutting” for decades, the artwork shows a rejuvenation state and thoroughly remoulds itself. In addition, the master also remakes the crown, thus creating a magic charm. Such technology of turning the foul and rotten into the rare and ethereal highlights the unique artistic charm of Lingnan Penjing.

The artwork is a tree, natural and ancient; however, it is also a vital sculpture, joyful and lively. The left and right trunks are like a couple holding, wandering and dancing together. Though in ripe old age, they still bloom the romance of life and bring the love enlightenment and the aesthetic education to the people.

真柏 *Juniperus chinensis* var. *sargentii* 陈国健藏品—— 2013中国盆景国家大展奖

Sargent Savin. Collector: Chen Guojian —2013 China National Penjing Exhibition Award

文：罗传忠 Author: Luo Chuanzhong

试想在高寒险峻的山巅峰口上，一棵被风霜雨雪摧折的柏树迎风而立。该树虽然全身伤痕累累，但经过若干年后，低托枝丛顽强生长，形成新的冠幅，且枝叶繁盛，生机勃然。大自然惨烈的打击没有扼杀它的生命，反而激发了生的希望，焕发了生命的第二个春天。这就是陈国健真柏作品给我们留下的直观印象。

作品最成功之处是把一棵直干树裁矮，并通过舍利干的技术处理，来达到“虽由人作，宛若天成”之目的。从外观上看，整树从根部到枝梢连皮带肉都被撕裂，重创处虽无法愈合，却形成了壮硕的水线，呈劫后余生的奇异景象。舍利处理雕刻技术娴熟，纽纹清晰，有规则随枝回旋扭动，如旋律跃动，余音袅袅。残留的枯枝随处可见，或残留于根部，或悬挂于树身，或掩埋于翠绿，或伸出于顶冠，平添了野趣，使人联想起大自然的鬼斧神工。

主要枝首集中在高位布托，此法非常考人，可见作者艺高胆大。上、中、下三大枝托皆向左倾斜，只留右枝作为衬托。底枝尽量下飘，增加动感；中枝穿插其间并向后横伸，增加厚度；上枝盘旋结顶，呈伞形圆顶。小枝疏密有致，繁而不乱，结构严谨而灵活，层次丰富而朦胧，且枝丛尽量向外伸展，层云叠翠，形成丰满完整的树相。

作品左放右收，整体走势向左垂探，外形就像蓄满劲道的弓箭，又似迎客的主人，既谦恭礼让又沉稳含蓄，让人百看不厌。

Trying to imagine that at the mouth of cold and precipitous mountain, a Juniperus destroyed by the wind, rain and snow stands upright against the wind. Though the tree is covered by wounds, after several years, the lower branches grow persistently and form a new crown with thriving leaves and vigorous vitality. The severe strike from the nature has not strangled its life, but arouses the hope of survival and glows as in the second spring of life. This is our direct impression of Chen Guojian's Juniperus chinensis var. sargentii.

The most success of the work is to cut the straight trunk tree short, and to reach the objective of "being handmade but as natural" by the technical treatment of Shari. From the out appearance, the whole tree, including the bark and the body, is tore from the root to the branches. Though the most injured part cannot be cured, it forms a robust water line, which appears as the wonderful view of survival after disaster. The carving skill of Shari treatment is consummate and forms clear twisted pattern which swirls and wriggles regularly along the branch as the leaping melody lingered in the air. The residual dead branch can be seen everywhere, or remaining at the root, or hanging on the tree body, or being buried under the green, or stretching out of the crown top. It has added the wild fun and made people think of the extraordinary skill of nature.

The major branches mainly concentrate at the high position. This skill is so difficult that we can see that the artist is highly skillful and generally bold. The three major branches at the upper, middle and lower positions all lean to the left, and the right branch has been saved as the contrast. The lower branch is made as floating downward as possible to increase the dynamic effect. The middle branch is made alternately crossing in it and horizontally stretching backward to increase the thickness. The upper branch spirals and merges at the top as the dome of umbrella. The density of limbs is in order, flourishing but not random. The structure is compact and flexible with rich and hazy levels. And the branches stretch outward as possible like a veil of clouds and Pinnacle and branches have formed the full and complete view of the tree.

真柏 *Juniperus chinensis* var. *sargentii* 高110cm 宽160cm 陈国健藏品
Sargent Savin. Height: 110cm, Width: 160cm. Collector: Chen Guojian

“盛世”五针松 *Pinus parviflora* 杨贵生藏品
——2013 中国盆景国家大展奖

“Flourishing Age”. Japan White Pine. Collector: Yang Guisheng
—2013 China National Penjing Exhibition Award

文: 徐昊 Author: Xu Hao

“盛世” 五针松 *Pinus parviflora* 高 110cm 宽 150cm 杨贵生藏品
"Flourishing Age". Japan White Pine. Height: 110cm, Width: 150cm. Collector: Yang Guisheng

松树能曲能直,受得沃土,耐得贫瘠。然其曲有劲节,直显本性,华而不失坚贞之气,清瘦更著青云之志,因此深得世人喜爱,成为能人志士托物寄情、借物言志的对象,唐朝诗人白居易更是“欲得朝朝见,阶前故种君”。自古以来,文人雅士皆好以松树制作盆景,置诸案头,骋怀味象,畅神其间,坐忘山林之乐。

杨贵生先生的五针松盆景“盛世”,寓情于景,具有丰富的人文内涵,令人观景生情,感悟其所表达的意境和内涵,从中得到美的享受,因此在首届中国盆景国家大展的评比中,受到评委和观众的好评,获得大展奖。

作品根基强悍劲健,主干呈曲折状,转折刚健有力,树干自三分之一处一分为二,形成双干,副干略细于主干,作同步转折,双干整体呈向上的气势,主、副干无论粗细、高低及势态,都极为协调而统一,给人以和谐之美。作者根据素材的特点，取低位出枝，枝势皆取略呈上扬的舒展状，布枝长短变化，疏密相间，错落有致，树冠结体如汉隶，厚重而朴茂，使之与树干的生理及形式非常协调。作者巧布枝势使原本矮壮的树相呈现向上的气势，因此使树姿显得宏大而充满蓬勃之气。

作品内动外静，内中充满生机，形式安定祥和，恰如盛年逢盛世，一派生机勃勃的和谐繁华景象，故名"盛世"。作者借作品的气韵气象，喻示当今经济发达、民生殷实、气象祥和、文化灿烂、社会和谐安定的盛世气象。盛世，是经济和人文的盛世，当今盆景文化的繁荣和发展，不正是盛世之象么!

The pine tree is able to bow and rise, enjoy the fertilized earth and endure the barren. However people love the pine tree because it bows with integrity and rise for its nature. It does not lose its faith during flourish age and stick to the high ambitions in poor conditions. Thus it has become the object for person of ideals and integrity to express their emotions and feelings. Bai Juyi, a poet of Tang Dynasty even wrote poetry as "I plant you right in front of the entrance step to be able to see you every morning." Since the ancient time, the refined scholars all like to fabricate Penjing with pine trees and place it on the desk to taste the view and have mind wandering in the joy of mountain.

The *Pinus Parviflora* Penjing "Flourish Age" of Mr. Yang Guisheng brings the emotion into the scene. It is full of humanistic connotation and stirs up people's feelings during their appreciation. The artistic conception and connotation provided by the work inspires people and brings them with joy of beauty. Thus, in the comparison of the first National Exhibition of China Penjing, it has been greatly appraised by the judges and audience and won the exhibition prize.

The root base of work is robust and powerful. The trunk appears in zigzag shape with firm bending. The trunk is divided into two trunks at the 1/3 position and forms the dual-trunk shape. The secondary trunk is a little thinner than the trunk and with synchronous turning. Both trunks are generally trending upright. The diameters, positions and gestures of both trunks are in extreme harmony and uniform, presenting people the beauty of harmony. According to the characteristics of material, the artist brings out the branch at the lower position. The gesture of branches is made spreading upright. The branch layout is variable in length and density and well arranged in order. The tree crown body is as Han script, which is thick, simple and thriving. It coordinates well with the physical style and manner of the trunk. The artist skillfully arranges the branches and makes the short and strong view of the tree to show an upright spirit. Therefore the tree gesture seems grand and vigorous.

The work appears in static and contains vitality inside. It is full of vigor but seems as calm and peaceful, just like the flourish age – vigorous prospect presented with peaceful scene, so named "Flourish Age". The artist has made a metaphor on the spirit of the work to express the prospect of developed economy, substantial livelihood of people, joyful atmosphere, splendid culture and harmonious and peaceful society. Flourish age is for the economy and humanity, and for the prosperity and development of the current Penjing culture. Is this the view of flourish age, isn't this!

**The work appears in static and contains vitality inside.
It is full of vigor but seems as calm and peaceful, just like the flourish age – vigorous prospect presented with peaceful scene, so named "Flourish Age".**

"大娇"真柏 *Juniperus chinensis var. sargentii* 陈伟藏品
——2013 中国盆景国家大展奖
"Delicate and Charming Girl of Heaven". Sargent Savin. Collector: Chen Wei
—2013 China National Penjing Exhibition Award

文: 张志刚 Author: Zhang Zhigang

盆景是"天人合一"的产物，是自然状态的树、石素材，经过作者精心选材、立意、培养、加工后，具有"人化自然"为特征表现的活的艺术品。欣赏盆景首先是感受作品的自然与活力之美，深层次的是领悟作品"二神"（自然之神和作者之神）一体、"二美"（自然美和艺术美）交融所表现出的艺术美和意境美。

"天娇"这件真柏盆景"虽由人作，宛若天成"，作品厚重大气，雄健豪放，尽显自然之风，毫无造作之匠气。给人的感觉是"苍劲、自然、健康、灵动、活泼、舒展"，可以称得上是一件完美之作。我虽不识其作者，但从作品中我能感受到他高深的艺术修为和过硬的加工技艺。

虽是台湾素材，揉合舍利加工，但作品却具有很强的民族特色，这与以往我们看到的很多日本和中国台湾的柏树在风格上有很大不同。该树为双干造型，一正一斜，一静一动，正者泰山压顶，疏密有致，动者斜干飘悬，灵动飞扬。在这幅险稳相依、动静相衡的画面下，很容易令人联想到生于危崖险坡之上的千年古柏，虽历经沧桑，饱尝风雨，仍岿然挺立，焕发着勃勃生机，她展现给人的不仅是美，更多的是一种力量，一种坚毅、不屈不挠的力量。

作品树干部分的艺术加工，是作品成功表现的关键之一。舍利干使树体变得苍老多韵而富于美感，通过色彩的对比和线条的变化，使原本圆滑的树干变得立体而沧桑。这种雕塑感很强的艺术感染力，只有柏树作品中才能体现。

"天娇"真柏 *Juniperus chinensis* var. *sargentii* 高 85cm 宽 156cm 陈伟藏品
"Delicate and Charming Girl of Heaven".
Sargent Savin. Height: 85cm, Width: 156cm. Collector: Chen Wei

"舍利干与神枝"一直以来是柏树盆景创作加工和表现的关键部分，她能增添树的沧桑感和美的表现力，但事物都是"一分为二"，如果作者艺术修为不够或技艺欠佳，对舍利部分的处理就会出现牵强附会、矫揉造作、画蛇添足、过于夸大表现等现象。而天娇的处理无论从宏观还是微观，整体还是局部，对于块面处理和线条变化都非常自然到位，分寸把握得很好。舍利展现也仅限于干体局部表现，并未夸大炫技，作品主体展现的仍是健康自然，积极向上。这一点是我对作品最大的认同。

作品所配莲花紫砂盆，无论从体量、款式，还是色彩都与作品相得益彰。但配座却有些牵强生硬，若换一自然边、平板座效果会更好。

Delicate and Charming Girl of Heaven

Penjing is the artwork produced by the "combination of the spirit and handwork". At first it is just the tree and stone material in the natural state. After the artists carefully select the materials, make up the conception, cultivate and process, it becomes the live artwork presenting the expression of "humanized nature". To appreciate the Penjing, we have to feel the beauty of nature and vitality of the artwork at the first, then further comprehend the "two spirits in one artwork" (the spirit of nature and the spirit of artist) and artistic beauty and conception beauty expressed by "the combination of two beauties (natural beauty and artistic beauty)."

Though this Sargent Savin Penjing "Delicate and Charming Girl of Heaven"is fabricated by man but as if made by the nature. The work is thick, grand, robust and wide, with full expression of natural style and no artificial traces. It is a perfect work bringing people with the feelings of "powerful, natural, healthy, ethereal, vigorous and stretching". I have not met the artist, but I can feel his artistic cultivation and excellent processing technique from the work.

Though the material is from Taiwan and combined with Shari treatment, the artwork is with strong national features, which is very different in the style from many Juniperus Penjing from Japan and Taiwan. This tree is of dual-trunk shape, with one straight and one sidelong, one static and one dynamic. The static one shows the weight as Tai Mountain with the density in order, and the dynamic one is with leaning branch hanging in the air and ethereally soaring upright. In the picture with combination of danger and steadiness and static and dynamic balance, it is easy to think of the ancient Sargent Savin standing on the dangerous cliff and steep, which stands still and glows with vitality after experiencing through the vicissitudes and hardships. It is not only the beauty she has presented to people, but more about the power, a firm and persevering power.

The artistic processing at the trunk part is one of the importance of the work's successful expression. The Shari turns the tree to be aged and full of beauty. The smooth trunk becomes solid and vicissitudinous by the color contrast and line changes. This artistic appeal in sculptural form can only be expressed in Sargent Savin works.

"Shari and the God branch" has always been the key part of Sargent Savin Penjing creation processing and expression. It can increase the historic sense and the appeal of beauty for the tree. But everything has its good and bad sides. If the artist is not good enough for his artistic cultivation or skill, the treatment to the Shari part will be far-fetched, artificial, superfluous and over expressed. The treatment of "Delicate and Charming Girl of Heaven" on the surface treatment and line changes is very natural and sufficient no matter from the macroscopic to microcosmic scale, or from the whole to the part. The limits are properly controlled. The Shari is also limited to the trunk local expression without exaggeration and flaunt. The main work expression is healthy, natural, active and positive. This is my greatest approval for the work.

The lotus Chinese clay pot complements with the work by the body size, style and color. However, the matched seat seems stiff. A naturally edged flat seat may bring with better effect.

In the picture with combination of danger and steadiness and static and dynamic balance, it is easy to think of the ancient Sargent Savin standing on the dangerous cliff and steep, which stands still and glows with vitality after experiencing through the vicissitudes and hardships. It is not only the beauty she has presented to people, but more about the power, a firm and persevering power.

畅谈中国鼎
—2013（古镇）中国盆景国家大展印象

The Impressions of China Ding-2013(Guzhen) China National Penjing Exhibition

古镇政府领导

何新煌 中山市古镇镇副镇长

古镇人种植花卉苗木已经有近40年历史，拥有深厚的花卉苗木产业基础。如今古镇的花木产业已经从过去个别农户的零星种植，发展为如今的"窗口+基地"、"公司+农户"集中大面积经营，一举成为中国南方最具规模的花卉苗木、园林绿化专业生产销售基地之一。

2008年举办南方绿博会后，中山的花卉苗木产业取得极为快速的发展。不过在发展的同时，古镇党委政府始终在考虑，如何进一步推动本镇的绿化苗木转型升级，毕竟古镇的土地比较少。盆景产业所占用的土地少，但未来相当有市场，而且已经逐渐走进寻常百姓家，完全可以作为古镇绿色产业转型升级的重点方向之一。

因此，古镇镇政府牵头搭台，举办了2012中国（古镇）盆景精品展。该次盆景精品展在"南方绿博园"展览馆举行。当时，来自韩国、越南、泰国、菲律宾等多国的盆景名家，以及盆景组织的高层都应邀出席盛会。精品展进一步提升了古镇盆景产业在业界的知名度。

2013年，首届中国盆景国家大展以及2013中国盆景艺术家协会会员精品展、2013中国古盆收藏展、2013中国盆景艺术家协会赏石展、2013中国盆景秋季贸易周等，均在古镇新建的多功能会展中心举办。这一展现中国盆景艺术最高水平的盛会在古镇举行，不仅对中山盆景爱好者来说是个福音，对于周边珠三角其他城市的盆景爱好者也是大喜讯。

首届中国盆景国家大展，无论是在参展作品还是在邀请嘉宾方面，均达到了国家级的水准。这100盆的盆景，是由中国盆景艺术家协会组成专家团从成千上万盆盆景精挑细选出来的顶级精品。为实现面对面交流，中国盆景艺术家协会还邀请了世界盆景友好联盟主席、欧洲、美国、亚洲盆景协会的多国会长，17个国家的盆景贵宾云集古镇。此外，为了激励广大盆景爱好者的创作热情，共同推动盆景产业的发展，展会不仅甄选出全国各省市知名的盆景团体及名家展品进行展出，而且评选出"中国鼎首席大奖"。为促进盆景事业发展贡献了全方位的推动力量。

古镇希望通过高规格的盆景展示、盆景文化交流、收藏品展示、行业交流和盆景交易，以及云集众多国内外盆景名流和业内人士，形成良好的展会品牌效应，使盆景产业如古镇的灯一样走入全球视野。

未来，古镇还将继续与中国盆景艺术家协会、广东省盆景协会合作，不断推动盆景精品交易平台建设，加快盆景产业文化发展，定期开展周期性盆景交易活动，逐步形成一个面向华南地区的盆景专业交易市场，为推动中国盆景文化走向世界作出一份贡献。

评委、监委

谢克英 中国盆景艺术家协会副会长 2013国家大展评比组组长

中国鼎——2013（古镇）中国盆景国家大展的成功举办，是中国盆景国家展览新的里程碑，为创建中国盆景国家大展展览品牌迈出了成功的第一步！实为可喜可贺！

中国鼎——2013（古镇）中国盆景国家大展整体策划方案，充分体现了中国传统文化和历史内涵，很有中国特色。“中国鼎”、“青花瓷图案”、“清明上河图”、“富春山居图”……一个个中国传统文化的符号向全球传播中国盆景的国家文化形象。策划方案改革创新，清晰时尚，有可持续性。

国家大展参展的作品是经过精心挑选的。从全国报名的数千上万的盆景作品中层层挑选出100盆参展作品，从展品的品种、质量、艺术水平等方面，应该还是基本反映了当代中国盆景的现状和发展水平的，应予肯定。遗憾的是有些松柏作品带铝线太多，成熟度不够。作品的代表性不够广泛。

这次国家大展的布展从10000m^{2}大展厅的选定、展台的制作、展品的摆布等都有所创新，与国内国际的同类展览有所不同，气派、高雅、清新。可惜的是有些展台的背景板不够宽，展厅的灯光不够亮。

国家大展从评比标准、评比方法、评奖方案等都是在认真总结经验和广泛听取意见后反复修改确定的，而且在宣传文件上、在展会上全部公开公布。

国家大展的评审委员会的监委、评委严格按照评比方案和评比规则对所有参展的作品进行了认真负责的评审。评委们对每盆参展作品的评分和评分结果都在展会现场公布，以供送展者和观众查询。我还建议将这届展览的所有评审文件、评分表及评审结果公开发表。作为评委负责人，我觉得我们评委会没有受到任何外来压力和干扰，也没有在哪方面搞什么平衡、权衡，而是在监委的全程监督下，评委各自独立地对每件作品进行评审打分。至于评审结果中有些获国家大展奖和会员精品展年度大奖、金奖的作品中带铝线太多等问题引起的反响较大。作为评委，对作品中带铝线的数量又如何界定等问题，都值得认真讨论研究，有待在今后的评比标准中进一步规范。对这次评比工作的全过程及存在问题，对不少盆景朋友网上或书面所反映的意见，在这里就不多说了，我会虚心广泛听取各方面的意见和建议，以后再找机会交流、沟通。

中国盆景年度晚宴很有创意，也很有新意，将颁奖、宣传盆景文化和娱乐有机地结合起来，这是很好的尝试。

评委、监委

魏绪珊 中国盆景艺术家协会名誉副会长 2013国家大展评比组监委

首届中国盆景国家大展，这本身就是一个吸引人们眼球和心灵的活动。改革开放以来，盆景事业的创作、发展也有一个翻天覆地的变化，由少数人的爱好到全社会的参与，并逐步发展成为一个产业，这是一个了不起的进步，首届全国盆景大展引起了全国及国际盆景界的高度重视。从参观者之众和国际众多友人的到来充分说明了这一点，这是一次让源于中国的盆景逐步走向世界的里程碑性的活动。

2013年，在山东组织了第六届鲁风盆景展，先后又参观了BCI扬州展和WBFF金坛展，扬州展展场的选择突出了古城风貌与盆景展的巧妙结合，将古城风景融入了展台及背景的设计；金坛展则采取室内外相结合的布展方式。以上展览各有长处和特点，均给人们留下了深刻的印象。首届中国盆景国家大展的命题就充分说明了它是对改革开放以来中国盆景艺术发展的一个阶段性的总结展，在场地的选择、布展的设计、展品的选送都做了精细的具体工作，300件展品多数都属精品，均能给人以耳目一新的感觉，它基本上展示了我们中国盆景人的敬业精神和力推盆景艺术走向世界的大家风范。同时，向世人展示了我们中华民族在盆景艺术上的文化内涵，一些外国友人的高度评价也充分说明了这一点。

本次大展给我的第一印象是进一步缩小了地域的差别，主要是对流派的进一步淡化，大部分作品都遵循了源于自然而又高于自然的这个基本法则。其次，在用材上多以松柏为主，这也符合世界盆景界的主流。就作品的高度提高到1.5m的问题应认真进行磋商方可形成规矩。

评比工作是展览活动中的一项重要工作，它能引导和推动中国盆景向更高水平发展，对评比的要求是公开、公正、公平，做到这一点的关键是要有一个德艺兼优的评委团队，要求这个团队的成员在业界要有尚好的口碑和对盆景艺术较高的专业知识，这样方能取得大家认可。这次采取打分的方法进行评比效果很好，减少了争议、偏见和一些消极因素，取得了较好的效果。在《中国盆景赏石》公布评分表是对公开、公正、公平的一个检验，同时，也是对每位评委的监督和评价。

评委、监委

王选民　中国盆景艺术家协会名誉副会长，2013国家大展评比组副组长

引人注目的“中国鼎——2013（古镇）中国盆景大展”圆满结束了，但在我的脑海里就像刚刚看完一部热闹而又动心的电视剧，至今心中还未平静。尤其是那个“年度之夜”，一时让人振奋鼓舞、一时催我深思，甚至激动了一阵儿，真的被感动了！当时我给在场的朋友说：国家大展成功了！中国盆景的历史将开始新的一页！“中国鼎”所产生的能量，将会进一步推动国内盆景的社会地位和盆景价值观的提升，对国际盆景会产生新的影响。

回顾一下展览，我认为入选展览的包括会员展的200盆作品一共300盆，展品还是高水平的，是近几年全国性盆景展览中好作品比例最高的！可能与本届提高展品入选尺度有关，当然也与展前作品选拔方法有直接关系，一份辛苦一份收获呀！当然事情也不尽完美，大家也会看到一些不那么令人满意的作品被选进展会，原因是在作品选拔的后期，只看照片不见实物，误差较大，所以以后的作品选拔方法要加以改进。展品的质量至关重要，能够入选大展的作品就标志着它的档次不同一般！说到参展作品我还是想旧题重谈，再次提出建议：希望协会尽快建立自己的“盆景创作委员会”！因为创作委员会能掌握协会具有创作实力的会员，同时也会掌握会员的创作档案，对于将来展览的作品选拔工作，提供展品资源大有益处。

这次大展的评比工作是大家最关注的事。本届所制定的评比规则和评比方法，总的来说是方法简便可以实行的。但是，在实践中我发现单纯的打分定论的方法还是存在一些问题，建议在以后的大展中还要加以改进。打分制要与“评”和“比”综合应用，打分结果出来后将前面若干名次列队进行实物对比，横向对比，也可以纵向对比。在评委主任的主持下，评委可以发表评论，进行对比，最终以表决或投票方式得出结果，经验证无误即可定论。盆景评比如果没有经过“评”也没有经过“对比”，仅是主观一票定高低，结果是不会完美的。我认为方法和手段不是目的，如能尽最大努力，用最合理的方法去得到最好的结果才是正确的，才能面对众人有所交待！事实上评比工作难的不是方法问题，而是评委队伍的组建，要求评委知识全面，具有丰富的创作经验和理论水平，还需要具备高度的责任感和艺德素质。有高水平高素质的评委队伍，才是评比成功的基本保障！应尽早建立自己的评审资格人才库，希望协会能重视并落实这项工作。

关于奖项分配问题，我认为盆景既然有形式分类，那么对参展作品也要实施奖项分类。如设大型、中型、小型、微型及山水和水旱奖项。同时制定合理配套的评比方法条例。这样做可以让各种盆景形式全面、平衡发展，以视公平对待。如何具体实施也许还需要一个过程，但面对当下盆景展览评比的改革需要，分类对待是必然要实施的，甚至不同分类的单项盆景展览的形式迟早也会出现，例如：大型盆景展、小型盆景展、山水及水旱盆景展、松柏类或杂木类展等。

以上仅为个人之见，意在给大家带来或引发新的思考，从而协力探索和推动以后盆景大展的良性发展。

最后，借此一角向为首届大展付出辛勤劳动的和支持大展的政府、企业单位及个人表示敬意！同时呼吁国内盆景同仁振作起来，目光转向“中国鼎”接受挑战吧！

畅谈中国鼎

—2013（古镇）中国盆景国家大展印象

The Impressions of China Ding-2013(Guzhen) China National Penjing Exhibition

评委、监委

徐昊 中国盆景艺术家协会副秘书长，2013国家大展评比组评委

本次国家大展吸引人的原因有两点，首先是“中国鼎——中国盆景国家大展”这个名称够响亮，够气派；再是作品的甄选方式，经过层层筛选和专家定审，使我感觉入选作品一定会够水准，够分量。

和近期其他一些展览相比，确实体现出国家大展的厚重感和时代性，这种办展思路是符合当今盆景事业发展潮流的。但作为中国国家品牌的盆景展，盆景的价值性和文化性要齐头并进，不能偏执于作品的经济价值而忽视盆景文化艺术性存在的意义。愚以为国家大展的入选作品应不分大小，以作品的水平为择要，让流行和时尚共聚一堂，这样才会赢得更广大的盆景人参与到活动中来，使国家大展拥有可持续性。本次中国盆景国家大展的布展设计品位较高，是前所未有的，唯灯光稍有欠缺，射灯照在立体的盆景上，形影叠加，影响视觉效果，可能以整体通亮的灯光为好。大展的展品总体水平较高，也是前所未有的，充分展示了中国盆景当前的水平和蓬勃发展的现状，其中也有几盆质量较差的盆景混迹其中，这在我们今后的办展中应当避免。

国家大展中确实有一些大而精的作品，令人震撼，而且艺术性和价值感并兼，符合当今社会的审美潮流。通过这次展览，我也感觉到大盆景确实价值不菲。200件会员展的作品中也有不少精品力作，充分体现中国盆景发展的势头和希望。50件古盆展品反映了中国盆景的历史发展轨迹，为展会增加了盆景的历史文化气氛。赏石与盆景具有异曲同工之妙，同堂展出，增加了展会的内容，同时也是一个相互交流学习的平台。

我觉得中国盆景艺术家协会的评比机制是目前较先进的方式，相较于其他评比中“议”的方式，这种方式不会受到某个评委先入为主的影响和干扰，能充分发挥每个评委对作品的客观判断而做出自己的决定，也避免了因个人的人情关系而影响到其他评委的客观判断，而且有监委全程监督评比，消除了评委之间相互为某件作品拉票之嫌。假如有评委因人情好恶，对某人的作品不切实际地给予过高分或过低分，在统计积分时也会成为无效分而被去掉。每个评委的评分表会在展览现场和书上同时对大众公布，这对于每个评委的业务素养和公正心是一个考验。如果评委的业务水平和职业操守都有问题，那么公布的评分表则是一面镜子，大家都可以在上面找到答案或进行质疑。所以我觉得目前的评比方式还是具有先进性和合理性的，至少是体现了公平、公正、公开的基本原则。

我多次参加这样的评比工作，以这种方式进行评比，每次的评比结果都还是比较满意的。虽然说和自己意想的得奖高低有些出入，但每个人的审美观不同，不可能都同一个想法，某个人的想法也不一定全对，有些出入自然是正常的。总而言之，优秀的作品都在得奖之列，大奖作品也在几件最好的作品之中，这说明评比结果还是比较客观公正的。

这次的中国盆景年度晚宴，是一个前所未有的盆景人的盛宴，可称开盆景历史之先河。我深深佩服苏放会长的创意和才华，他将现代光影和历史文化相结合，融汇到盆景艺术中来，使之产生既震撼又优雅的视听效果，提升了盆景艺术的品位，可谓盆景之幸，盆景人之幸矣。其中的古筝演奏尤其高雅入境，古筝音乐无形的线条，就像盆景的线条那样跌宕起伏，动人心弦。在晚宴上，我被授予“盆景理论文章年度奖”，我深感荣幸，也感到惭愧，泱泱大中华，万千盆景人，人才济济，岂唯我不敏之辈？苏放会长的文章就比我好，他的文章宏观全局，有情，有理，其中不鲜盆景观，看似荒诞不经，细想却如醒醐灌顶，令我惭愧之至。

我有幸参加了在西安唐苑举办的“世界盆景论坛”，并作了专题演讲。我觉得这是一个东西方盆景学术交流的大舞台，也是文化交流的平台，通过这种形式的交流，使我加深对西方盆景文化的了解，同时也是向外国朋友宣传中国盆景文化的大好机会。我是首次参加这样的国际论坛，通过这次活动，使我认识到今后要以怎样的方式来演示，才能更好地让西方人理解中国盆景文化，达到传播中国盆景文化的目的和效果。对于我来说，这是一次学习的机会，在此也感谢张小斌董事长为这次会议的付出和盛情款待。我希望下次论坛的主题是“盆景的人文精神”，因为任何艺术都具有其民族的人文精神，同时也希望有更多的中国盆景人参与这样的活动。

我觉得本次中国盆景国家大展的筹备工作做得有条不紊，各项工作都做得很好，使大展取得圆满成功。每次在广东举办大型展会的组织工作都做得很成功，广东的盆景人办展会是有能力、有经验的。

作为一个盆景人，首届中国盆景国家大展对我是有积极作用的，它使我看到了中国盆景更加美好的前途和希望，增强了对盆景事业的信心。

中国盆景艺术家协会 会长团队

柯成昆 中国盆景艺术家协会常务副会长

古镇是个具有岭南特色、欧陆风格的生态城镇，是人杰地灵之所在，2013（古镇）中国盆景国家大展在此地召开，吸引人的不仅仅是盆景艺术精髓，还有对中国的非物质文化的认可。本次国家大展汇聚了国内外著名盆景大师、专家及全球盆景界知名人士，是中国近代盆景史上首次由国家一级盆景协会主办的国家级专业水准大展，是一次最高规格、最高水平的中国盆景艺术集体汇演，堪称盆景界前所未有的一次新的里程碑，中国盆景新时代即将拉开帷幕，具有重大的历史意义。相信大家也是一样翘首期盼，和我一样带着同样欣喜和期待的心情前往大展。

大展的活动突破了以往的组织形式，不仅结合中国盆景国家大展、中国盆景艺术家协会会员展、中国古盆收藏展、中国盆景艺术家协会会员赏石展等几项类别大展，还增添中国盆景年度之夜颁奖晚会，设立了多项奖项颁奖、技艺表演和文化交流活动，推出了作为中国盆景国家文化品牌象征的“中国鼎”为名称的中国盆景首席大奖。活动多样，给大家营造了一个欢娱轻松的氛围，同时也展现了国内外盆景人士风情万种、多才多艺的一面。这些设计令人刮目相看，极具创新，这样高规格的组织模式在历届盆景展中都是绝无仅有的。

“中国鼎”是以具有中国传统文化和历史内涵的中国商代鼎造型为模版，通体采用全玉材质精心制作而成，价值不菲。但此次首席大奖的评选设置醉翁之意不在酒，意义不在于奖，而在于一种激励，鼓舞。激励盆景爱好者的创作热情，鼓舞大家共同推动中国盆景产业以及中国传统文化的发展。这是中国盆景艺术家协会以及各省市盆景协会盆景人士共同努力和期望的方向。

此次大展的另一个看点，是评委团队的组成及相应的评比机制。对于每一位送展者而言，获奖的意义不在于奖杯奖金，而在于奖杯背后所寄托的分量，那一份比金银更可贵的肯定和精神支持。当大家都把自己精工细作的作品送到展会上来时，对我们的评委就是十分的信任和期望，所以评委组能做的，就是真正的公平、公正和公开，以更加透明的方式对待评奖工作，用更加严格的规则赢得送展者的信赖。这是我们今后各项盆景大展应当借鉴和发扬的一种评比体制。

此次活动组织比较紧凑，各项安排也十分到位，一些细节上的工作比较细致，如接待、住宿等，组委会也煞费心思，做了大量的策划和组织工作，尽可能做到全方位的服务，这种精神和态度是值得肯定的。当然，我们也期待下一届的国家大展能够更上一层楼，一届比一届更加精彩更加振奋人心。

中国盆景艺术家协会 会长团队

黎德坚 中国盆景艺术家协会副会长

在与朋友学习、交流盆景的时候，对一个盆景人来说，印象最深的还是这个作品给我带来了什么？这次参展作品在规格上可以说是达到了最顶级，我怀着极大的兴趣想看到中国各个流派在"最大体量"上是如何表现盆景艺术的。

我也曾参加过不少展览，对这次大展的评价可概括为精炼、浓缩、大气、提升。我们在细看时，可发现许多作品的细节绽露出艺术的创意，从整体看，感觉本次展览就是全国盆景的微缩，就是中国盆景的代表。大展再次体现了苏放及团队的工作辛劳和卓越，我们看到，每次国庆他都带队奔波在外，尤其是他和盆景友人共同提出的"盆景只有友谊没有国界"的理念更成为每次大展的亮点。来自20个国家和地区的盆景友人在大展现场热切地交流切磋，他们一路走过广州、中山、深圳，来到东莞"真趣园"共聚晚餐，切磋盆景技艺、巩固深厚友情。在园林花卉的多项产业上，好像没有哪个项目如盆景这般有如此强大的国际凝聚力，这一切都是苏放会长的重要贡献。

这次的参展作品普遍都有桩材老、盆龄长、桩型沧桑等特点。看来参加选拔、送展的作品都经过精益求精的筛选，体现了主办方的心血和期待。

通过这次展览，我感到我国盆景不但流派众多，而且每个流派中个人的创造又在不断地丰富、传承和改变着自己所立足的艺术根基。

中国盆景年度晚宴富有创造性、时代感。例如水鼓舞、女模盆景秀以及舞台的灯光、音乐、节奏、总体制作等都令人耳目一新。

大展的筹备工作有序，确认书及活动通知等都非常及时、准确、到位、人性化，通知后留有充裕时间；但对于活动宣传，本人认为可能是经费问题，所以在宣传力度和知晓程度方面稍有欠缺。我认为在时间紧迫、召集不便的条件下，可以让参展者或大会组织者提供电子版展品图片，使评委以浏览的形式先"热身"，再初选时评委对参选作品的把握或许会更好。

在接待、送展和撤展过程中，我觉得大会组织工作总体是成功的，但要做好各个环节不容易，所以还是有不便和不畅通的时候，例如送展和摆放时就显得乱，如果现在还达不到用电脑控制整个流程，就可以招募志愿者在最紧张的时候来帮忙，广州亚运会召集志愿者的做法可以借鉴。

这次总的评比结果是令人满意的，评比的方式也无可厚非，"三公"的程度也很高。但评比事关重大，万万不可粗心大意，我们还要重视评比的方法，例如我在前文提出的评委先"热身"的方法，即评委要先熟悉参评作品，初选后还可以"差额选举"的方法公布出来，让更多的专家和高手甚至会员参与。有时我们强调防止"故意的不公"，但对"难免的不公"还要采取有效的措施。也就是评委要有"备"而评，避免草草了事。此外评比工作现在可以逐步规范化、标准化了，就像体育大赛很多本来不可量化的项目也可以对照标准来打分一样。这只是我个人的点滴想法，请大家参考。

作为众多盆景人中的一员，对于盆景未来的发展，正如我前面所提到的：它给我带来了什么。我会为此积极思考、学习和尝试新作。总之，艺术不能失去个性，个体的特点是艺术的生命；但艺术也不能失去自然的和人文的环境，环境是艺术的源泉。

中国盆景艺术家协会 会长团队

李城 中国盆景艺术家协会副会长

很荣幸参加了2013（古镇）中国盆景国家大展。这次展览最吸引我的地方是所有展品都是高质量、高水平的盆景。要做到这点，其实是非常难的。这次是组委会的专家们亲自跑到全国各地看盆景实物，初选出盆景后再照照片，又通过照片综合比较才最后选出参展作品。这样做虽然耗时耗力，但它保证了展品质量与中国盆景国家大展的名实相符，也会让广大会员形成"能参加中国盆景国家大展就是荣耀"的共识。这个共识非常重要，它会缓解盆景人仅仅把参展获奖作为目标的局限性，而把目标扩大为能够参加展览，这样也就能让更多的人有更大的热情投入到盆景事业中来。当然，我们不能每次展览都这样挑选展品，这只是一个良好的开端，一个榜样。我相信，有了这个开端和榜样，今后各地挑选盆景送展的组织者们也会像这次组委会的专家一样，严格把关，认真挑选！

中国盆景艺术家协会 会长团队

李伟 中国盆景艺术家协会副会长

中国盆景随着中国政治、经济社会的改革走上了一条振兴之路，回顾中国盆景30年的发展历程，特别是收藏家群体融入盆景圈以后所带来的一些盆景观念的变革，确实对当今的盆景艺术注入了勃勃生机，迎来了从来未有的飞速发展。有了坚实的物质基础而由此派生了一系列相关政府及收藏家牵头的大展。无疑为广大盆景人提供了展示盆景艺术的良好平台，这个平台既展现了中国盆景的独特魅力与技艺、鉴赏方面的交流，同时会更有力地促进盆景艺术的普及和技术的提高。国家大展的推进已经为未来的中国盆景巅峰之作踏平了道路。

艺术的追求是无止境的，盆景是一种永远无法定型的特殊艺术，盆景的创作会随着一棵树的不断生长变化而灵活的变化。创作者也会随着制作技能及审美情趣的提高而对一些盆景作品进行不同程度的改作。所以这些特殊性因素就决定了盆景会不断产生新的变化。

当今社会正处在变革时期，众多确定的、不确定的因素都会对人们的物质及精神需求产生一些新的影响和变化，盆景也不例外。中国盆景正处在一个变革时期，一方面域外盆景在树种选择、成熟度及养护技术方面已经超越了我们很多，另一方面我们在如何继承传统如何创新的理念上尚未形成成熟的理论及创作思潮。盆景收藏、创作、经营等方面还有一些盲目跟风、明星效应等误区。因此，我们盆景人必须时刻保持清醒的头脑，正视盆景现实环境中存在的问题。只有这样我们才能少走弯路，与时俱进。

作为中国盆景人的一员，我们有幸赶上了改革开放的好时代，由于中国经济的飞速发展，为中国盆景的复兴踏实了物质基础，口袋里有钱了，才能激活盆景收藏的环境。盆景国家大展顺应时境，为盆景人摇旗呐喊，搭建了一个上档次、重品牌、理念新的全新平台，这是中国盆景人的骄傲。在国家大展的影响下必将掀起盆景收藏的超级风暴。回顾历史，展望未来，任重道远。通过首届国家大展的展示我深有感触。第一，参展盆景作品应忍受残酷的时间代价。盆景作品一定要先具备较高的艺术价值，再考虑其经济价值。当然好的作品可以肯定的是两者皆优的，大展上有不少这样的艺术精品。三尺盆盎中首先展示的是一棵小中见大、古树态的树，可以肯定的是它的整体状态一定是苍老但健康的；它的根干枝叶的造型、分布是合理的，有利于在比例上更优地体现小中见大的效果，且具有良好的传世性。造型上符合美学原理，空间有合理的布局，气韵贯通，符合自然法则造型技法。这些要素就决定了必须付出时间的代价，以及金钱的代价。第二，重视文化内含。中国盆景的特点就是具备诗情画意，情神兼备。强调景、盆、架三位一体。盆面布局合理、陈设环境讲究、展览环境与人文气息相得宜彰。第三，尽快建立盆景人才的培训、培养机制。人才是一切事业发展的决定因素，高素质创作人才的缺失是中国盆景发展的瓶颈，通过多种形式的交流及培训迅速培训一批高水平的创作队伍及素质全面、技能扎实的理论研究人才已成当务之急。

中国盆景艺术家协会 会长团队

李晓 中国盆景艺术家协会名誉常务副会长

此次“国家大展”吸引我的地方有两点：1.中国盆景艺术家协会能两次（2012年、2013年）在古镇举办盆景展，说明了地理位置的重要性；2.展品是中国盆景艺术家协会通过专家们半年以来在全国精心筛选而来的。

此次展览启用“国家大展”的名称，体现了中国盆景艺术家协会在同类协会中的地位。本届“国家大展”在办展思路，布展设计，组织等方面都有很好的新意。通过这次展览、总结经验、不断地完善，把“国家大展”办成世界级的品牌展，同样此次2013（古镇）中国盆景国家大展的“高水平、多样化”，体现了组织者的能力和“中国盆景”的实力。

这次国家大展和会员展300盆的参展作品，总体水平很高，盆景制作技法的呈现各有新意，能把国内南北的盆景技法及世界各国的盆景技法、内涵、古盆、赏石融为一体，真是一个百花齐放的“国家大展。”这次“国家大展”发现“岭南树种”制作的技法，能南北相融，所呈现的意境效果极佳，像此次获得“中国鼎”首席大奖的作品，定意明确，飘逸有致，技艺精巧，是一盆上好的作品。

我觉得本次“国家大展”的评比规则和评比机制是当前最好的方式，评委们也公平、公正的对待每位精心选送的作品，对于将在《中国盆景赏石》中公布评分表的举措，提高了“国家大展”的公平、公正透明度。我对评比结果是比较满意的，在本届“国家大展”所有参展的作品都很优秀，各具特色，奇崛苍劲、或横逸斜，希望大家取长补短，来年拿出更好的参展作品。我对本届“国家大展”的筹备工作很满意，组委会也很辛苦，评委也很公平、公正，感谢他们的辛勤周到的工作，我希望组委会在以后的展会活动前及时通过多种方式联系我们！

我个人认为中国盆景年度之夜是中国盆景艺术规格最高、水平最高的一次集体“汇演”，更充分地体现了展会的主题，所以我认为这次晚宴是很成功的。

“趣怡园”是吴成发先生历经20年艰辛打造而成的岭南盆景创作和展示基地，为中国盆景走向世界做出了贡献；“真趣园”黎德坚副会长敢于新的尝试，并矢志不渝培育出的“真趣松”的精神对我的启发很大，受益匪浅！西安唐苑举办的“世界盆景对话论坛”推动了盆景艺术的普及、提高，很感谢张小斌先生的精心安排，“唐苑”主人张小斌先生不仅给西安市民提供了一道丰盛的文化大餐，更为众多来自全国各地的盆景爱好者的观摩、交流提供了一个难得的平台！“唐苑”是一个多文化以服务为一体的都市文化生态园，它体现了自然景观与人文景观的有机融合！

通过此次“国家大展”，作为中国盆景人的一员，我对盆景艺术也有了进一步的认识，在此次参展活动中学到了很多的园艺盆景知识，同时也认识了不少盆景界的前辈，当然也希望自己在以后的盆景制作方面不断地创新和尝试，多学习交流，共同推进中国盆景艺术的创新与发展！

中国盆景艺术家协会 会长团队

储朝辉 中国盆景艺术家协会名誉副会长

2013（古镇）中国盆景国家大展，是代表国内最高水平的盆景展览，还有大师现场制作表演，是国内外盆景界明星云集的地方。本次国家大展，档次高、形式新颖，从入选作品、展台设计都体现了高质量、高水平，我个人认为能和日本的“国风展”相媲美，展区布景高端大气有内涵，打造出了中国盆景人的梦想和展览品牌。精彩的晚宴盛会提升了盆景艺术行业的含金量，也为中国盆景艺术走向世界开创了先河，以前没有经历过，这是非常好的开端，用不同形式表现，让盆景人和盆景艺术互动起来。

首届中国盆景国家大展的展览展品都是精品，尤其是100盆国家大展作品，造型美而不失自然、枝条老道、布局经典、诗情画意、内涵丰富，给人留下了深刻的印象和回味的空间，但各地区的盆景水平差异还是存在的。从本届国家大展活动中可以看到，本次办展思路、中国国家品牌展的设想、布展设计、展品入选方式，都体现了中国盆景艺术家协会在以推动中国盆景艺术事业发展为主导的创新精神和团队力量。这是实现盆景艺术中国梦的希望！值得一提的是本次大展的接待服务热情周道，尤其是组委会电话跟踪安排交通路线、住宿、活动日程等相关事宜。

我个人认为中国盆景国家大展这种由9位评委和监委组成评委团队的打分制评比方式很好，这种评比方式同时也是对评委专家素养的一种要求和考验，只要评委们不是会同打分，最后统计也能体现出评委水平的差异。在《中国盆景赏石》里公布评分表，还有这样的打分制挺好，公平、公正、透明。我对这种评比方式非常满意，原因有二：①打分制，体现了公平、公正；②张榜公布每位评委的评分，杜绝了人为暗箱操作。

中国盆景年度晚宴新颖、有创意，用商业色彩包装盆景。盆景的商业价值一直在被开发，这次晚宴是个很好的开始，晚宴有好几个环节令人印象深刻：模特展示盆景、沙画、最后阶段各国盆景友人一起跳舞，这样的晚宴给人感觉像“影星”颁奖一样。盆景展览是行业人的聚会，盆景晚宴也可以成为文化潮流的趋势，盆景人期待的盛宴。盆景商业化运作也就更自然了。

深圳的“趣怡园”，是一处宁静优雅的所在地。这里的盆景更加让人心旷神怡，数以千计的盆景作品，风姿绰约，千姿百态。游走其间仿佛进入盆景“大观园”。园主吴成发先生是中国盆景界的一位杰出的大师和实干家，他创作与改作的盆景作品佳作数量之多，且形式多样，种类丰富，其中既有树木盆景，又有水旱盆景；既有观叶类，又有观花、观果类，每盆都年功老道，展现了岭南盆景风姿，妙趣横生。让人感受到中国私家盆景园的特色和实力！

东莞的“真趣园”是一个有浓郁岭南特色的私家盆景园，全园景色简洁古朴，落落大方，以自然为美，却又整体布局严谨。园内的山水配置十分考究，园中有园，景外有景。种植的大部分是海岛罗汉松，以及九里香、台湾松、日本松、本地山松等，极具观赏价值的盆景桩材多。造型各异的精品盆景、地景高低错落、分布有致，充满了诗情画意。听了相关介绍，知道园主有着强大的经济实力和推动中国盆景艺术发展的心愿，令人敬佩之至。

作为一名盆景艺术爱好者，参加本次活动备受鼓舞，同时也开拓了视野，在心中也产生一种压力和期待，期待自己更加努力提高，把事业做得更好。

中国盆景艺术家协会 会长团队

罗贵明 中国盆景艺术家协会副会长

2013（古镇）中国盆景国家大展是中国盆景发展的一大进步和改革，是规模较大、发展较快、质量上乘的一次国际性、研讨性展览。

本次展览在组织和形式上都有新的亮点，特别是与圈内其他的展览相比，不仅展场的布置更完美、细致，展品也较多，质量上也是精品层出不穷，更为创新的是把盆景文化与人文文化有机地结合在一起，使人耳目一新，让盆景艺术得到了升华，让人与自然更有机、更紧密地相互增辉。

我对这次展览的印象很深刻，创作形式新颖的作品辈出，所采用的材料也有新的突破。展品中精品太多，无论是创意上还是技术手法上都有提高，比如有些中小型作品更注重的是创意的内涵性和抽象派艺术的展现。有的作品更是传神达意地展现了其内涵，如黄杨作品“揽月”。

颁奖晚宴的确很有新意，是本次大展的一大亮点和特色，这就是把盆景艺术和现代人文艺术有机结合得最完美的部分，特别是将盆景、会徽、会旗与T台艺术相融合，更彰显了盆景艺术的尊贵和高雅。

大展前的筹备工作开展得很好，也很周详，只是关于后续活动的组织最好能应统一行动。组委会的接待工作很周到，工作人员很辛苦，非常感谢他们!

我对大展的评比结果应该是比较满意的。大展评奖的评分规则和机制都很不错，也是目前较好的方式。建议今后可随机抽取一些观众和会员进入评委组参与打分，让群众分在评分中占一定比例，这样会更公平、公正。在《中国盆景赏石》中公布评分表的做法很好，更能增加透明度。

本届盆景国家大展的成功举办和创新，使中国盆景产业的发展得到了很大的推动，同时也让中国盆景的价值得到了提升，在创作技艺和立意上有了新的定位和理解，这会让盆景创作者们注重文化与盆景的结合。

中国盆景艺术家协会 会长团队

芮新华 中国盆景艺术家协会副会长

2013（古镇）中国盆景国家大展的展品是通过专家们耗时半年，从全国筛选出来的一流的作品，从最开始就让人充满期待。

中国盆景国家大展的名称比较恰当，布展、设计等思路都很有新意，希望通过这次展览总结经验，不断完善，把中国盆景国家大展办成中国国家品牌展。展览的内容总体能体现中国盆景的实力，展品多样化，南北盆景技艺得到了很好的交流，并取得成绩。

大展和会员展的300盆参展作品总体水平较高，盆景制作技法也有新的创意，是一个百花齐放、世界各国盆景技法融合的展示。今后参展作品还可多样化，范围还可更广一点。

这次荣获中国鼎大奖的作品是岭南树种，按岭南盆景制作技法制作、达到较好意境效果。关山先生的作品“飞天”也有不凡之处，首先定义明确，疏密关系处理得当，飘逸部分恰到好处，丝雕适度、技艺精巧、配盆得体，算得上是一盆完美的好作品。

我对中国盆景年度晚宴感受是惊喜，惊的是古老的盆景艺术也能同现代舞台多元艺术融合一体，既满足了现代人的眼欲，又体现出晚会主题。喜的是今后盆景晚宴有了更多创新，内容更加丰富多彩。这次晚宴非常成功，模特盆景走秀是个亮点，我很喜欢。

组委会的筹备工作做得很好，只是活动通知应尽量提早。我参加这次活动一切都很顺利，组委会很辛苦，谢谢他们辛勤周到的工作。

这次评比规则和评比机制是当前最好的方式，评分表的公布增加了评审的透明度，我支持。希望评委们为获大奖的作品做一个点评。我对评比结果总体上是满意的。得奖作品不一定百分百优秀，没得奖的作品不一定就不优秀，好作品太多，奖额有限，能参展已经很荣幸了。另外就是大、中、小类别盆景应有一个比例。

参观深圳“趣怡园”、东莞“真趣园”非常愉快。首先感谢二位园主的热情招待，二位园主对盆景的热爱、追求、执著、勤劳的精神鼓舞着中国盆景人，中国盆景大有希望，中国盆景的发展离不开他们。个人建议把盆景展示区和养护区分开，这样展示效果会更好。再次谢谢二位园主。

西安唐苑举办的世界盆景对话论坛太好太及时了！它让世界更深入地了解了中国盆景，也让中国了解了世界盆景。希望下次论坛主题是《盆景文化与技法》，这需要提前约稿、翻译好发言稿，节约时间。这次西安唐苑的张小斌先生为论坛提供了一切会议条件，花费了大量财力、物力，感谢张总的热情款待。

通过这次活动，我对盆景有了进一步认识，今后在制作方面要多元化尝试，创新方面多下工夫，成功与否不重要，重要的是尝试。

中国盆景艺术家协会 会长团队

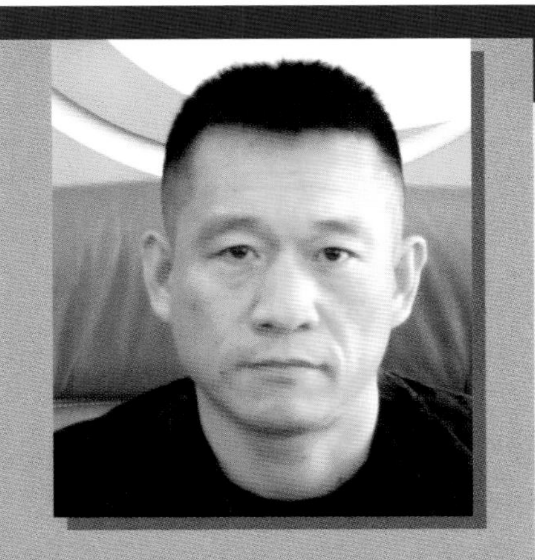

于建涛 中国盆景艺术家协会副会长

2013（古镇）中国盆景国家大展是一个有创意、有规模、有气势的全国性大展。相比中国台湾的“亚太展”和日本的“国风展”，我们这次展览有质量、有档次，可以说为中国盆景走向世界又跨出了一大步，也提升了中国盆景艺术家协会的品质与形象。我认为只有中国人才有如此“大手笔”，所以要做出我们的“中国味道”。

我觉得目前的评比规则和机制还可以更完美，我们也可以吸取中国台湾、日本等大展览的成功经验，借鉴他们的评比方式和打分规则，结合中国的实际情况，制定一个越来越好的评比细则。

中国盆景年度之夜的晚宴相当成功，我觉得这是整个展览活动中最有意义、最具创新的环节！晚安的现场气氛浓烈，效果绝佳。其中沙画节目结合了盆景艺术、传统文化，形式新颖，内容丰富。特别是晚会最后，苏会长亲自上台带领大家热舞，把晚会推向高潮。

本次大展的筹备工作还算不错，工作人员非常辛苦，感谢他们的付出。

我作为中国盆景圈人中的新人，对于盆景制作、欣赏还在学习阶段。关于收藏，通过这次展览，我看到一个契机，就是“盆景拍卖”，四年前我曾说过如果盆景能像古董字画一样拿到拍卖会上去拍卖就好了，现在看到我的这个设想终于实现了。我相信，将来盆景的价值、升值空间一定会超过古董字画。

送展人

2013（古镇）中国盆景国家大展期间全国各地盆艺高手云集于一地，互相切磋技艺，并开设盆景交易区，让展览与交易同时进行，是两个吸引人的地方。而相对于国内其他展览，2013（古镇）中国盆景国家大展选择的地点——古镇非常理想，当地气候条件能满足南方和北方盆景的养护需求，这也是此次展览的展品水平高的原因之一。种盆景的人都爱盆景，试问气候不适合谁敢把好作品送来呢？这次展览的布展背景和灯光布置可以说是我所见国内展览历届最好的，没有之一。相对于国外展览此次展览主题具有强烈的中国风气息，“问鼎”不仅让人感受到中国古典文化中的风雅情怀，同时又将此情怀转化成新时代盆景人追求的中国梦，它向世界展示出中国盆景的独特风格、魅力，以及中国盆景人的高雅品位。遗憾的是会员展的展台只有60cm高，宽度也感觉不够（目测），有不少作品的枝托贴住了背景布。如果展台高80~90cm，盆景作品

陈万均 中国盆景高级技师
中国鼎——2013（古镇）中国盆景国家大展送展人

的枝托离开背景布10cm以上会更有利于作品的艺术展示。

这次展览的展品具有较高的艺术水准，云集了国内盆景的高档精品，形式多样，印象最深的就是中国鼎上的盆景，经典隽永，把中国最有代表性的盆景作品置于中国鼎上是大会给予该作品和作者最大的荣耀，中国鼎之巅代表中国盆景盛会中的最高荣誉，它激励着每位盆景人，去寻找梦想，开拓创新，创造出更多、更好、更精彩艳绝的中国风盆景。古盆与赏石在展览中起到了补充的作用，古盆是中国古典盆景文化中的重要部分，形制简朴但经典耐看，让人细细品鉴之后仍久久回味。

中国盆景年度晚宴最精彩的部分是盆景走秀，这种表现形式在很大程度上提高了盆景的表现力、可塑性和发展潜能，扩充了盆景欣赏的角度和方法，让盆景在视觉呈现上得到更新的表现，是中国古典风雅与西方的时尚艺术品位的结合，从艺术价值的角度来讲，这次晚宴让盆景的发展迈进了一大步。如果走秀的模特在穿着、打扮上能多添加一点中国元素的话会更加精彩。

接到大会的通知书和邀请函时，我的第一反应是要先细细欣赏一下封面的精致设计，然后再慢慢阅读里面的内容，一张用心制作的邀请函是对参展者的尊重，这一点大会做得非常好。我的盆景初选是由广东省盆景协会会长曾安昌先生负责，他亲自来到我盆景园内帮我甄选作品参展，并讲解参展规定与评比标准，这尽心尽力的工作态度令人敬佩。同时，由于我的送展路程比较短，所以送展和撤展的过程都非常顺利，没有遇到任何困难。

于现在来说，2013（古镇）中国盆景国家大展的评比规则和评比机制相对来说是公平的，我认为是到目前为止最好的评比方式，在《中国盆景赏石》里公布评分表，也符合公平、公开、公正的原则。我认为，对于评比结果我觉得不能说满意和不满意，换个说法是相信和不相信或者公正和不公正。已然要参加此次展览就要在相信组委会和评选小组的前提下才会来参加。已然来了就是信任，信任了就不会质疑。在这里我有一点小小的建议，在展品多样化方面希望可以作更具体的区分，增加微型盆景区和超大型盆景区，让展览的内容更为丰富多彩，体现多元互动、和而不同、融会贯通的展览精神，让国家大展不仅成为业界的权威盆景盛会，同时也能助力那些具有新思想，热衷新形式的盆景艺术家获得业界的认可，进一步促进当代盆景的发展。

中国国土辽阔，地理环境和地域文化也相差甚大，这导致了产生出各地风格不一的盆景创作技法。2013（古镇）中国盆景国家大展和同期的会员展让全国300多盆优秀的盆景作品汇于一地，加快了中国南北盆景文化的交融与碰撞，在这样的环境下盆景创作人互相学习、开拓思维、加强意识，相信以后会有更多优秀的盆景因此次展览而诞生。国家大展对盆景的品鉴和收藏都具有导向作用，吸引各地藏家聚集，推动高档精品盆景的收藏价值，深化盆景文化内涵，建立含盆景展览、养护、收藏、品鉴于一体、超越产业发展的良性生态圈，这是我对未来的盆景盛会的愿景。

送展人

王鲁晓 中国鼎——2013（古镇）中国盆景国家大展送展人

2013（古镇）中国盆景国家大展吸引我的首先是海纳百川的胸怀，它不分疆土、不分地域、不拘一格选精品，只要是好作品都有机会登台展示。其次是消除了“一言堂”、“这个地区我说了算”、“师徒父子全包揽”的现象。

这次展览给我的突出感受就是主题鲜明，宗旨明确。与之前各类展览相比，办展思路新颖，品牌设计独具匠心，布局设计和展品水平空前，整体展品的严谨性、成熟度凸显。此次国家大展最大看点是精品繁多、质量上乘，公开、公平、公正是另一大看点。

本届大展筹备工作井然有序、有条不紊，而在活动宣传方面，我认为适当与政府互动一下，如：对入选国展100盆的作品，组委会及时将入选情况通报给各地政府并邀请其参加国家大展这一活动，通过双方互动，进一步让各级政府认识到盆景是中国传统艺术的重要组成部分，从而引起他们的高度关注，促进地方盆景发展。其次是要进一步严格对盆景初选工作，如参选杂木盆景，应以“寒枝”图照作为筛选的首要与必要条件，因为只有通过“寒枝”才能展现作品盆龄及枝条的过渡、成熟情况，把优秀的作品展现给公众，真正体现竞争公平、程序公开、评选公正。

从以下两点再次印证此次大展是一次公开、公平、公正的国家大展：一是在国展和会员展中，排名较后的不乏盆景界知名人士，而一些名不见经传的作者走向了奖台；二是对没有按要求的作品全部挡在参评大门之外，这是评审团队对国家大展的高度负责，是对所有参展人员和观众的高度负责，真正体现了严格标准，不讲情面。同时为更加公平和透明，我建议在评审结束后，评委和被评人员进行互动，即让参展人员对评委、监委也进行打分考量，实行末位淘汰，让那些德艺双馨，严格、公正的评委更具公信力，继续留在评审团，为下届评选提供强有力保障。

中国盆景年度之夜是一个盛大空前的晚宴，它直接为中外创作人员提供了交流、宣传的平台，对中国盆景传统文化的输出、传播将起到积极的作用，可谓耳目一新。在整个颁奖过程中，特别是对总冠军及8个金奖作品，如果请专家逐一点评，让每一名获奖者发表个人感言，我想晚宴将更加充实。

由于整个活动工作量较大，建议在下一届展会中，进一步加强组织管理，尤其在布展和撤展过程中相应增加工作人员，以使活动善始善终。

首届中国盆景国家大展的成功举办，将对延续和传承中华文脉、展现盆景的魅力和生命力具有崭新意义和重大影响。

送展人

杨鹏勇　中国鼎——2013（古镇）中国盆景国家大展送展人

这次参加中国盆景国家大展，我感到很高兴。与近年来的展览相比也是相当不错的，相对满意的。

在经济允许的条件下，独立的展台更独特，更能展现盆景的魅力，也是这次展览的一个优点。

但是，100盆大展的作品水平也是参差有别的，和最好的盆景相比，有的差距还比较大，而会员展中的作品也不乏极好的。100盆中国盆景国家大展的作品等级高于中国盆景艺术家协会举办的历届全国展金奖一个等级，如果不是最好的，或者其中有质量偏低的，便不能服众。如果有的盆景还没有会员展中优秀的好，就会引起争议。这就要求我们在筹备选拔时，严格把关，不分参选时间先后，只要确实优秀，就可以随时调整，以保证100盆的高质量。

此外，不提供运费补贴会给送展带来一定不便，但是我们通过同一地区几个人一起送展克服了这样的困难。

送展人

2013（古镇）中国盆景国家大展高规格，高水平，集中了国内大家最顶级的作品。通过本次的展览，我感觉到中国的盆景展品水平已经和国际接轨，与扬州展的水平相当。

展品方面：这次参赛的100盆代表了中国盆景造诣方面的最高水平，与以前相比有了很大的提高，200盆会员精品与以往相比有较大的创新，逐步与国际水平接近。值得注意的是，各派别的盆景创作理念差别较大，应该注重各派的交流，吸取他人长处，才能使中国盆景走向世界。

筹备方面：筹备工作比较细致周到，挑选程序比较合理。广东入选部分是由广东省盆景协会会长曾安昌带队从基层挑选，比较切合实际，也能发现有创意的好作品，这一点予以保留。

评比方面：评比比以往更加公开公正，比较完善。

颁奖晚宴方面：参加的外宾比较多，节目比较丰富。

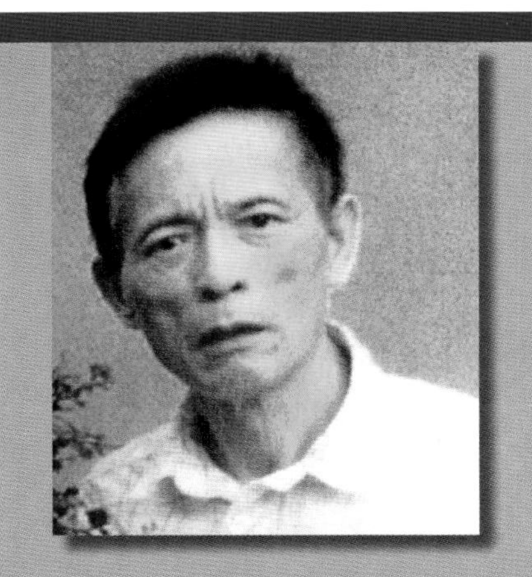

尤广才　湛江盆景协会会长　中国鼎——2013（古镇）中国盆景国家大展送展人

送展人

朱昌圣 中国鼎——2013（古镇）中国盆景国家大展送展人

2013（古镇）中国盆景国家大展的送展区域便利以及展区周围氛围活跃都是吸引我参展的原因。

首届盆景国家大展的所有展览展品（100盆国家大展、200盆会员精品盆景及46件古盆精品、52件观赏石精品）都代表了送展者的精心付出，最深刻的印象是中国盆景有如此氛围，如此多用心为行业付诸努力的盆友。所有盆友都在寻找新的理念表达方式，其中一件关山收藏的罗汉松"苍翠"展品体现了一种全新的概念，在《中国盆景赏石·2013-9》里面有叙述其理念的创新概要。

中国盆景年度颁奖晚宴让人感觉新奇，我本人最喜欢的部分是水晶鼓舞表演，但最牵动人心的部分应该还是颁奖仪式。

可以看出组委会为本届大展的展前筹备工作花了很多心思，但在执行方面可以更细心高效，比如酒店住宿、进展安排等。

中国盆景国家大展营造的氛围让全国很多盆友聚集此处，可以达到很好的正面交流的作用，交流可以带来很多不同区域的看法以展现大家不同的理念方式。

我遇到酒店住宿安排的问题，送展人被安排成随行人员。总之，组委会的努力大家有目共睹，辛苦了！

送展人

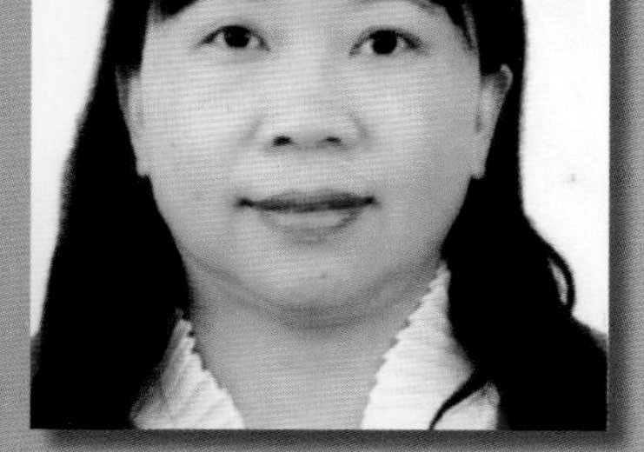

邓秀珍 中国鼎——2013（古镇）中国盆景国家大展送展人

2013（古镇）中国盆景国家大展是中国近代盆景史上首次由国家一级盆景协会主办的国家级专业水准大展，它的开幕预示着向全球盆景界宣告中国盆景新时代的开始。

这次展览是由中国盆景国家大展、中国盆景艺术家协会会员展、2013中国古盆收藏展、2013中国盆景艺术家协会会员赏石展以及中国盆景年度之夜颁奖晚会等几个板块组成。汇聚了国内外著名盆景大师、专家及全球盆景界知名人士，开展盆景制作技艺表演和文化交流活动。

通过参加首届盆景国家大展，欣赏到诸多十分出色的盆景作品及古盆和赏石，让我大开眼界。

我觉得中国盆景年度晚宴是新鲜的、精彩的、无与伦比的。这个颁奖晚宴的价值在于我们之间能得到更好地交流。

这次大展的筹备工作都比较到位，在接待、送展和撤展过程中，我没有遇到什么困难，很顺利，希望今后的展览继续延续这些优点。

关于本届大展由9位评委和监委组成评委团队，我希望今后的评比机制可以是专业评委和大众评委共同投票。大展的评比结果我是满意的。

最后，通过参加本次中国盆景国家大展，我深刻意识到中国盆景同样是与时俱进的。

畅谈中国鼎

—2013（古镇）中国盆景国家大展印象

The Impressions of China Ding-2013(Guzhen) China National Penjing Exhibition

送展人

陈再米 中国盆景高级技师 浙江省盆景艺术大师 中国鼎——2013（古镇）中国盆景国家大展送展人

首先，我很高兴参加2013（古镇）中国盆景国家大展并且作品入选中国盆景国家大展100盆顶级盆景，此次大展最吸引我的地方有如下五点：1.参加此次大展是每个盆景人最大的荣誉，也是盆景人特别珍惜的盆景梦；2.本次大展是目前国内级别最高并设有最高奖金（中国鼎）的展览，如运动员不想当冠军，就不是好运动员；3.大展云集国内外顶级高手，是高档盆景集中展示、交流、观摩、学习最难得的机会；4.参加此次大展使我进一步了解国内外盆景的市场行情，与自己创作盆景艺术水准处在什么档次定位作对比；5.在展会上，相会老朋友，结识新朋友，增进友谊。

其次，这次2013（古镇）中国盆景国家大展，自从盆景作品初选开始，苏放和鲍世骐两位会长在酷热的天气亲自到全国各地盆景园挑选精品佳作，让我感受到本次大展同以往的展览大不相同。这是咱们中国盆景艺术家协会高层团队非常重视的战略举措。盆景作品经层层筛选把关，到确认100盆入选作品，再到公正、公开打分是其他国内展所没有的，可以说这次是特别严格地走上了规范之路。这是一次叫得响的中国国家品牌大展，这么大的展厅、宽敞的空间，高档次顶级的盆景作品，充分突出“苍老、宏大、粗干”的中国盆景特点。精品盆景成熟度高和优美的构图，不管是松柏类、杂木类、花果类、山水、水旱类等堪称是国内超一流之作品，也是我所参加过有史以来档次最高的展览。中国国家大展品牌名称响亮，战略思路高端、独立式大展台、展品顶级高档以及中国文化元素都在这次活动中淋漓尽致地体现。

此次首届（古镇）中国盆景国家大展，我送展的盆景里，有2盆盆景荣幸地入选，另有2盆榆树参加会员精品展。纵观100盆国家大展，参展作品造型独特、个性十足、大气磅礴、年功老到、构图完美、苍古入画，充分展示了当代中国盆景无穷的艺术魅力和深厚的文化底蕴，那“源于自然，高于自然”的艺术视觉冲击力，在我心中留下深刻印象。如容桂盆协罗崇辉先生选送的九里香盆景，作品高128cm，表现了一株在祖国大地上生机勃勃、奋发向上及岁月留痕的沧桑感！看那抓地有力、树相古朴、枯骨舍利反卷黑白分明，枝托定位精准，争让得体、主次分明、苍劲有力、多头结顶、枝脉清晰。右边的大枝托似在迎接远方的朋友到来，如没有天长地久的精心呵护，难有佳作呈现。

会员精品展中也不乏佳作，有些是够资格入选国家大展，只是在体量上略小“1号”，整体造型上正在缩小差距。因为创作者、收藏者，把自己认为最好最满意的盆景，送上最高档次的展台来展示，这也是可以理解的。自实行打分制评比，本人创作的榔榆“万众一心壮中华”能得到大师们中肯、客观、公开、公正评判与定位很满意，使我找出差距迎头赶超。

就古盆、赏石作品展而言，每位收藏家无私奉献展品以供人们学习鉴赏之用，不仅借此平台向世人展示了世界一流的中国赏石、古盆文化，也更好地全方位地呈现了本届大展多样化的中国元素。

随后，感谢中国盆景艺术家协会高层团队精心策划首届国家大展和整个中国盆景年度晚宴，真是不容易！盆景人仿佛回到娘家一样，格外亲切。这个晚宴始终充满热闹、喜庆、激动的气氛，令我难忘！另外，本届大展中外嘉宾齐聚一堂，不仅拉近了中外盆景友人间的距离，而且增进大家的友谊，互通信息，其乐融融。我尤其喜欢的节目是带有中国盆景文化元素的“沙画”表演和“模特盆景秀”。

本届大展展前活动通知及宣传初选程序等在下一届应保留，只是在挑选盆景作品上还要精益求精，更上一层楼。本次大展有那么3~5盆不尽如人意，最后希望以后的展览展品宁缺毋滥。

整个大展在接待、送展、撤展过程中和展前筹备都做到有条不紊、环环相扣、一丝不苟、认真细致，帮助我们排忧解难，包括安保志愿服务人员及工作人员等都尽心尽力。特别值得一提的是东道主陆志伟大师、王金荣高级技师负责布展、接待等任劳任怨、认真细心、极其负责，还有无数叫不上名来的志愿工作者，为大展做出无私奉献，精神可嘉，令我由衷赞赏，感慨万千！希望在下届送撤展时，给予较远省份晚一天送达的宽裕点的时间，撤展时尽早一天离开，以便节约时间和不必要的开支费用。

心底无私天地宽，在《中国盆景赏石》里公布评分表，将是对每一位送展者尊重且激励与鞭策的举措，使参展者从中找出差距和不足，作品放在展台上，让数据来说明，是再好不过了，俗话说硬碰硬，让人心服口服。这一举措本人将举双手赞成，我想大多数人也是赞成的。

实际上早在几十年前，本人就树立了超强精品意识，在选桩制作、收藏、欣赏等方面，按制定方针去作精品盆景。杂木为主，松柏、梅桩盆景、花果类并举，走自己的自然式造型特色之路。时刻关注国内外盆景的发展方向，紧跟时代潮流。“众人拾柴火焰高”，每个盆景人精心创作出符合时代主旋律的优美佳作，来迎接中国盆景文化强国梦的到来！

送展人

阮建成 2013中国盆景艺术家协会会员精品展送展人

作为中国盆景人，盆景展对他们的吸引力无疑是巨大的，特别是2013（古镇）中国盆景国家大展。因为通过参展可以知道和了解现时中国盆景创作、创新水平的最新动向、信息。通过参展可以对自己的作品、藏品作一次高水平的评价，展现自我作品、藏品的最佳风采。最好能获奖，得到盆景界的肯定，会对自己以后的盆景创作产生积极的影响。

2013（古镇）中国盆景国家大展与（BCI）扬州展、（WBFF）金坛展，室内和室外举办展览存在本质的区别（室内受地方、空间等的局限，室外则有大自然的风景衬托，场地空间大、光线自然、空气清新等优点）我认为室内展览应多增加一些人工景点的素材（如小桥流水、假石山草地、亭台楼阁等）进行点缀，会使参观者心旷神怡、悠然自得地欣赏盆景的优美风采。

2013（古镇）中国盆景国家大展在名称、办展思路、中国国家品牌展的思路、布展设计上都着重突出了“中国鼎”元素。中心思想贯串了整个展场，使参观者将“中国鼎”深深记入脑海，不易忘掉，比其他国内展览领先一筹。本届展品的整体水平档次都有较大的提升进步，体现于展品的成熟度、整体的造型气势、枝托的比例协调等方面。

2013（古镇）中国盆景国家大展上使用的灯光（一盆一灯）营造出了良好的盆景感观，突出了盆景的层次，强烈的光线对比，使盆景树的枝、干显得更加苍劲有力，叶色显得更加青翠，更显健康活力。

从盆景创新理念方面来看，这次展览也有不少好的作品，如国家大展区038号的罗汉松，在360°各个角度都能得到良好的体现，不存在正面、侧面、背面的区分。这对盆景及作者创作都提出了更高的要求，在创作中，出枝的走向取舍都要充分考虑各方向的因素。

本届的中国盆景年度晚宴可谓星光熠熠，专家云集。中外盆景名人欢聚一堂，交流盆景的心得及发展方向，有很深远的意义，也是对中国国家大展举办单位、参展者肯定的体现，把本届中国国家大展提高到了一个新的高度。能为中国国家盆景事业出一点力，作为盆景参展者我感到无尚的光荣与自豪。

我最喜欢中国盆景年度晚宴的颁奖环节。中国盆景艺术家协会的领导成员全体上台亮相，体现了本届协会领导的团结协作精神。

2013（古镇）中国盆景国家大展由9位评委和监委组成评委团队，是目前为止最佳的评选方式（奥运会跳水项目都用这一方式），去除最高分和最低分后，剩下几个评委分数的相加除以剩下评委数，就得到该作品、藏品的总分。再者，在《中国盆景赏石》里公布评分表，对参赛者实为公开、公正、公平。公开评委的姓名、评定作品、藏品的分数，即使评得有误差，也能反映该评委的专业水平如何，让大家一目了然。在公开、公正、公平的方式下，作为评委一定会认真、负责的。我对此次评选结果是非常满意的。

作为众多中国盆景人中的一员，2013（古镇）中国盆景国家大展给我未来的盆景制作、收藏、欣赏都起了促进作用。先是吸收先进的理念，再进行自我消化，不断进行实践和理论学习，多看别人的作品，多结交盆景友人并与之进行交流学习，也希望今后自己能有更多的好作品、藏品能参加中国盆景国家大展。

嘉宾

李添津 中华盆栽艺术台湾总会理事长

2013年中国台湾组团到大陆参观盆景展，有4月中旬于扬州举行的BCI国际盆景赏石大展、9月24日常州金坛的亚太盆栽展以及9月29日在广州中山古镇中国盆景艺术家协会所主办的国家大展。

古镇的国家大展在室内展示，空间大，台上每件作品都独树一格，皆是主角，质与量都是空前的，可见主事者的用心。开幕晚宴有多位气质高雅的模特儿手捧小品盆栽走秀，令人赞赏，印象深刻。

盆景审美由配盆到主角，树龄越苍劲越美，越大越有雄伟气魄，有时候残缺是岁月摧残的另一种美的表现。至于大盆栽（约120cm），套一句老话“数大就是美”，譬如我们看到的大工程、大机械、大豪宅、大钻石……大又精致绝对稀有可贵，以我个人也较偏向喜好收藏此类作品。

中国台湾地处亚热带，杂木类盆栽树种较多，虽南北距离只有400多千米长，但全年度阳光、温湿度、海拔高低及雨季都有明显的差异性，所以选择合适树种很重要。俗语说：“成功是优点的发挥，失败是缺点的累积”，管理盆栽一定要先了解各树种的习性，经验丰富的玩家懂得筛选土壤，浇水的动作是功夫，适时的修剪、施肥、除虫害以及换盆、剪除饱满的须根或剔除老叶子让它再生都是学问。

中国台湾的“华风展”在台北花博争艳馆举行，室内空间大，有湿度喷雾辅助，参展作品由顾问及企业家提供，而所有参赛者皆由各地方县市盆栽协会年度比赛得奖作品代表参加，有一定水平，成熟度够。在大陆所看到的唯一缺憾是有少许作品若晚个2年或3年再上架将会更完美。

中西碰撞 艳惊寰宇
——国际著名盆景人眼中的“中国鼎”

【美国】威廉·尼古拉斯·瓦拉瓦尼斯
《国际盆栽》杂志 出版人兼主编
[America] William N. Valavanis Nicholas Publisher and Editor for *International BONSAI* Magazine

我被2013中国盆景国家大展所征服。在其他任何地方我从没看到过如此漂亮的盆景。与我在中国看到的盆景相比，美国和其他国家的盆景不能公平地称作“盆景”。中国盆景的尺寸很特别给我留下十分深刻的印象，但以我对盆景的研究，不用看展牌介绍，我能猜到这些树的尺寸到底有多大。展场里的树都很优秀、与众不同，不仅具有动态美而且修剪得很到位。起初我想知道盆景展台之间的过道为什么那么宽，后来仔细想了下终于明白了，机械或卡车需要足够的空间来回运输这么巨大而沉重的树。

在本届大展上能看到中国古盆和赏石展，令我惊喜万分。展厅后半部的中国盆景艺术家协会会员展与100盆国家大展的树形成了鲜明的对比。这些树更像世界盆景的规格，整个中国国家盆景大展和各种各样的盆景树种给我留下深刻的印象！

中国盆景的造型与我见过的、研究过的和教授过的盆景的造型有很大差别，大多数盆景更有雕塑艺术美感，强调高而纤细的树干而不是树的特性，一些看上去像异想天开正在翩翩起舞的人们。这些元素大部分都不同于过去50年里我一直研究的传统日本盆栽。

与日本盆栽和水石相比，我发现中国盆景与中国赏石的理念差不多。例如，日本水石是小而硬的，黑色调为主，表面呈水平状，给人以安宁的感受。中国赏石更趋于大而柔，光亮又以浅色调为主，表面呈垂直状且赋予变化。大部分传统的日本盆栽的规格都比中国盆景的规格小，且突出三角形轮廓给人以安静祥和的感觉。

起初，在中国看到这么多大尺寸的盆景令我震惊，但是现在我理解这是为什么了。不像日本，中国是一个广阔繁荣的国家而且有巨大的花园。在日本，人们有在室内壁龛摆放盆景的传统，所以它们必须是小体积的。在中国，花园很大而且我觉得人们没有在室内摆放盆景这么强烈的传统意识，因此，中国盆景通常被摆放在室外园林。如果是更小体积的盆景，像日本尺寸的盆栽放到那么大的园子里也许会被弄丢，而且不会像放在中国的大园林里那样给人留下深刻的印象。当日本盆栽被摆放在室内时，更具欣赏性，而中国盆景通常被陈列在大园林里，只有放在那里，中国的大盆景才会显得比例协调。

参观中国盆景大展时，我注意到大师们使用同日本差不多的修剪技法而且这种技法遍及全球。剪枝、蟠扎和塑型等技法都被运用在这棵树上。关于中日两国如何整理树姿这方面，我看到没有什么差异。但是我确实注意到一种独特的岭南派盆景风格，这类盆景的造型没有使用绑扎的手法，但是树枝也同样造型良好。这点令我很惊喜，但在国外我没见到有充足的关于岭南盆景的研究实例来扩充我的知识。

在本届大展上我仔细观察了两盆盆景的现场制作表演。我看到唯一的差别是现场的语言障碍，我不懂中文，所以听不懂中文的描述，提问和回答。与日本的制作表演相比，当然我也不懂日语，但我认为不需要太多的语言交流，我只是通过观看表演者的现场表演从中学习。

我很高兴能被邀请和参与如此重要的盛事。主办方和组委会确实做了一项杰出的工作。我组织过30多场座谈会和大型会议也知道这种大会涉及多少繁重的工作，但每个人都出色地做好了本职工作。

另外，就我个人而言，由于我摔坏的脚所带来的不便，我真心感激给予过我细致入微帮助的人们，使我不落队跟上团队的行进速度。现在我的脚已经痊愈，我期盼着将来再次来到中国继续我的盆景研究。

这趟中国参观访问的经历让我大开眼界！这次我经历到的盆景是我之前没有想到过的。它们漂亮，不像在美国通常可以见到的那些。很愉悦看到中国在推广和发展盆景艺术和盆景艺术起源这

Penjing Culture Clash Between China and the West Surprise the World

—— Famous International Penjing People's Views on "China Ding"

条道路上或许起到了带头的作用。因为几十年以前在美国，我真没听说过也没见过如此优质大气的盆景。

神奇的中国盆景使我的盆景经历更为精彩，我期盼将来到中国在盆景起源国继续我的教学。最后，通过融合不同的形式、历史和各自关于盆景的理念，我们能提升盆景和盆栽的制作水平并传遍全球。

小贴士：

由中国盆景艺术家协会设计兼具古典与时尚美的桌旗，成为展场上一道亮丽的风景线，受到中外嘉宾的赞许。

I was "blown away" at the First 2013 China National Penjing Exhibition! I have never seen such beautiful Penjing anywhere. The Penjing I've seen in China, the United States and other countries do not do justice to the word "Penjing." The size was quite impressive, but I can see through the size in my study of Penjing. The trees were outstanding, distinctive, dynamic as well as well trained. At first I wondered why the aisles between the Penjing were so wide, but later thought about it and figured out machines or trucks were necessary to move the huge, heavy trees.

It was a pleasant surprise to see the ancient Chinese containers as well as the viewing stones. The secondary exhibition by the China Penjing Artists Association was a sharp contrast with the large size trees. Those trees were more like the sizes of bonsai around the world. I was truly impressed with the entire exhibition and variety of plant material!

The shape of China Penjing is quite different than the bonsai I've seen, studied and taught. Most of the Penjing are more sculptural and artistic, emphasizing tall slender trunks rather than the characteristics of the tree. Many look like people dancing and are whimsical. These elements are mostly different from the classical Japanese bonsai I've been studying for 50 years.

I find the China Penjing is similar to China viewing stones, when compared to Japanese bonsai and suiseki. For example, Japanese suiseki are small, hard, dark, horizontal and quiet. Chinese viewing stones tend to be large, soft, light colored, vertical and busy. Classical Japanese bonsai are mostly smaller than China Penjing and present quiet feelings with prominent triangular silhouettes.

At first I was shocked to see so many large size Penjings in China, but now understand the reasons why. China is a large prosperous country with gardens, unlike Japan, China is bigger. In Japan there is a tradition of displaying bonsai indoors in tokonoma, so they must be small. In China, the gardens are larger and I don't think there is a strong tradition of indoor display. But, rather, China Penjing is normally displayed in outdoor gardens. Smaller, Japanese size bonsai would be lost and not impressive in Chinese gardens. The Japanese bonsai are more appropriate when displayed indoors, while China Penjing is generally displayed in large gardens where they are in scale.

When viewing the China Penjing I noticed the same training techniques used in Japan and throughout the world. Pruning, wiring and shaping are used to train the trees. I see no differences between how the two were trained. But I did notice a style of Penjing where wire was not used, but rather, the branches were tied for shaping. This interested me but I did not see enough examples to study the techniques to widen my knowledge.

I carefully watched the two Penjing demonstrations at the exhibition. The only difference I saw was that I could not understand the Chinese language descriptions, questions and answers. Of course, I can't also understand Japanese either, but I don't think it's necessary too much. Just by watching the demonstrators I learn.

I felt honored to have been invited and included in such an important event. The hosts and organizers did a superb job. Having organized over 30 symposia and conventions I know how much work is involved and everyone did an especially good job.

Also, personally speaking, I truly appreciated all the considerations and assistance given to me because of my broken foot which provided me with the opportunity to keep up with the others. My foot has now healed and I look forward to returning to China in the future to continue my Penjing study.

My trip to China was an eye opener for me! The Penjing I experienced were nothing like the concept I had before. They were beautiful, not like what is commonly available in the United States. It was a pleasure to see China take part and perhaps lead the way for promotion and progress in the art of Penjing, the origin of the bonsai art. For decades we have not really heard or seen quality Penjing in the United States.

My entire experience was a wonderful introduction to the fascinating world of China Penjing and I look forward to continuing my education in the origin of bonsai in China in the future. Together, by combining different forms, history and concepts of artistic potted plants together we can raise the level of Penjing and bonsai throughout the world.

【意大利】玛利亚·基亚拉·帕德里奇 BCI（国际盆景俱乐部）理事

[Italy] Maria Chiara Padrini
BCI Director

在我参加过的所有中国盆景展中，这次的中国盆景国家大展是三大盆景展之一。毫无疑问，这次的盆景展胜过我之前参与过的所有中国盆景展。展览的举办场地很宽阔，无疑给了人们全方位欣赏盆景之美的机会，使盆景成为一件艺术品。在赏石展中，尽管大部分都属于河石，但是却带给人们赏心悦目的感觉，让我们有这样一次机会可以尽情地观赏这些珍贵的石头。精美的木质底座更加凸显了石头的精致。我十分赞赏在整个展览过程中对于展台和背景的独特设计以及后续的维护，使得展品与展场融为一体，带来更好的展出效果。

从西方盆景与中国盆景的不同之处来说的话，西方国家更多的受到来自日本美学的影响。与此同时，在数十年中，日本人逐渐把栽培技术转变成为一种艺术表现形式，把树看成有生命的材料。对于形式美的疯狂追求，使得日本人更加执著于技艺而不是树木本身的自我表达。这种趋向无疑造就了技艺精湛的盆景人，除了少数盆景大师以外，这种趋向有时也阻碍了树木本身最自然的表达，削弱了树木的个性。中国盆景形态的构思，较少的涉及雕刻手法，使得更加自然地表现树木本身。在西方人眼中，这样的盆景看上去也许会显得比较乱和不那么精致，但是同时也表达出一种蕴藏在树木本身的活力以及能量。

通常外国人认为中国盆景很大，就我本身而言，我自己有不一样的感触。通常认为中国盆景很“大”的人，也许从未到过中国。中国是一个很大的国家，必须要考虑到它的国土面积。我注意到一些盆景是用来装饰一个很大的空间，它们虽然很大，但是放在特定情境下小的几乎看不见。如果我们说起美学，就要提到和谐以及平衡。必须要从全局角度想问题而不是把注意力单单放在一个方面上。

关于盆景技法方面的不同，我认为其实国家间的盆景技法没有特别大的不同，通常都是判断一棵树的习性然后慢慢改造它，直到达到最终的审美目标。真正有区别的是国家间不同的哲学思想和审美观。

国家间的盆景艺术必须进行交流才可以共同发展。我认为，不同文化之间要找到一个可以促进盆景创作和盆景展示的方面来进行交流。当然，这种交流必须要考虑到因不同国家各自的历史以及哲学因素导致的审美视觉的差异。随之而来，加深对影响其他国家艺术发源的因素的理解。没有文化，又何谈艺术。

我尤其想对2013（古镇）中国盆景国家大展的组织者们说一声感谢，干得好！

对于中国我有一种特殊的感情这是我第八次来到中国，第一次是在2005年。中国以惊人的速度向前发展，每次都像是到了一个新的国家。当你觉得你已经足够了解中国的时候，突然意识到事实上对中国一无所知。我在中国遇到的每个人都很为别人着想，充分考虑我的需求。伟大的国家拥有伟大的人民，就当我马上就要坐上载我回家的飞机上，不想离开中国的想法愈发强烈。

This is the third largest exhibition in China in which I participated. Undoubtedly it is better than the previous ones. The exhibition set in large areas gave the chance to admire the specimens in all their beauty, promoting them to works of art. The stone exhibition, although mainly represented by River Stones, was really good and we could appreciate precious stones. The quality of the wooden bases with refined workmanship emphasized the excellence of exposed rocks. I enjoyed a greater care and refinement throughout the exhibition, both in tables and backdrops that in the pots and stands more suited to bonsai trees exhibited.

Western countries have been very influenced by Japanese aesthetics. This, in the same time, during some decades, Japanese transformed what was a cultivation technique in an artistic expression using the tree as a living material. The frantic quest for perfection in form led to the search for more rigorous execution of techniques than the expression of the tree itself. This trend has undoubtedly created excellent technicians, but with the exception of the great masters, has sometimes overshadowed the natural expression of trees, reducing their personality. The way of conceiving of the Chinese Penjing, less sculpted and rigidly set, allows a greater naturalness. In the eyes of a Western, Penjing may seem messier and less refined, but this conveys a greater vitality and energy that lives in living things.

Western people tend to think that China Penjing is large.Those who think that the Chinese Penjing is "large" perhaps have never been to China. China is a big country and must be considered this size scale. I noticed some Penjing exposed to adorn a very large room. They were big bonsai but nevertheless almost disappeared while being there to represent the theme of the meeting. If we talk about aesthetics we refer to harmony and balance. They must carry out considering the whole and not a single element.

Concerning the difference of techniques between different countries I think that techniques in Penjing are not so different between countries, always it comes to forming a tree knowing his reactions and adapting them to the aesthetic goal we wish to achieve. What are different are the aesthetic and philosophical schools of thought.

I think there should be an exchange between different cultures to find each other those aspects that can improve the way to create and display a bonsai (Penjing). This exchange must also concern the historical and philosophical aspects that have different aesthetic visual. Consequently to deepen the understanding of the factors that has influenced the art in the countries of origin. Without culture there is no art.

I would say especially thanks to the organizers, well done!

This is my eighth trip to China, the first in 2005. China is growing at a rapid pace and always it seems to me to visit a new country. When you think you know enough you realize to know almost nothing. The people I meet were caring and took my needs into consideration very fully. Great people in a great country.... and the wish to get back right away grows when you just walk to the plane that brings you back home.

【日本】 须藤雨伯
景道家元二世
[Japan] Uhaku Sudo Keido Iemoto (headmaster) II

这次展览的盆景作品都是令人钦佩之作，让我学习颇多。而且，我对岭南盆景的理念有了更多了解，特别是岭南作品要经过长年累月地修剪，这一点与日本盆栽不同，实在可佩。对于水石，中国赏石、奇石、山水赏石等是在与日本水石有别的种种价值观之中成立的，我们应该从现在开始学习中国赏石。对于盆器，日本的名器都是缘自中国的名器，我觉得没有什么差别。而今后深入地研究盆器的历史和传统是十分必要的。

对于中国的大型盆栽来说，我没有感受到任何不适。由于对环境和价值追求的世界观是不同的，因此中国盆栽之大让人感到充满品位与风格的活力，是可以接受的。

谈到中国盆景与世界盆景，可以认为日本文化的审美观、理念、意境、风格、风韵、技术等是日本盆栽的摹写范本。但是，日本盆栽的精神层面却是禅的美学与心境，并且是独立的审美意识。闲寂、幽静是日本文化的象征，而其具象化的作品就是盆栽。因此对盆栽的关注就相当于对日本文化的关注，和日本审美意识达成共识。我认为世界盆栽是在日本盆栽、日本文化的基础上成立的。世界盆栽和中国盆景的不同即是中日文化的不同，到最后将是不存在差别的。重要的是对别国文化的尊重、理解，世界盆栽文化的发展将是具有各个国家特征的盆栽共同存在。另外，日本盆栽与中国盆景都有存在的意义，没有必要统一他们的称呼。

正因为文化、习惯、传统的不同，世界的盆栽才妙趣横生。我期望彼此都具有各自的特征，同时，认为必须要了解盆栽/盆景的本质。盆栽是从中国传来的，随着时间的流逝盆栽与盆景的文化、风俗、习惯迥然不同。可以说，是盆景艺术在日本的土壤上通过日本人的审美意识培养而发展形成的日本盆栽艺术。盆景未来的发展将是从某处起源，并在那个国家的文化与习惯中培育，诞生具有各国特征的盆栽/盆景。我认为重要的是存在差异。

现场示范是在短时间内学习有限的技术，具有很强“秀”的要素。日本也是参考美国的这种形式而举办活动，但这种活动并没有诠释原本的盆栽的存在状态。原本的盆栽技术，并非人做，而是通过自然的力量与时间的累积形成的自然造化的事物，人的技艺只是有限的。由于演示时间有限，就要求产生立竿见影的效果，这与原本的创作盆景或创作盆栽是背道而驰的。一直以来，示范表演都存在问题。如果可以演示出具有本国特征，又独具一格的作品，将非常有趣。希望今后研究出以学习中国盆景本质为目的、追求“天人合一”、“意境”、“气韵”境界的示范表演形式。

中国盆景的发展势头迅猛。而且，最令人欣喜的是，年轻的盆景制作家、盆景人满怀热情，更多的人关注盆景并努力学习盆景技术。在中国，盆景方面优秀的专家这样多，着实难得。中国盆景正阔步走向未来，这是盆景人的幸事！

All the exhibited Penjing artworks, from which I have learnt a lot, are admirable. And I have understood more about the concept of Lingnan Penjing. Especially the artwork has to be trimmed for years, and that is very different from Japanese bonsai and really impressive. For suiseki, Chinese scholars' rocks, rare stones and landscape scholars' rocks are founded in various value views different from the suiseki. We shall start learning Chinese scholars' rock since now. For pots, all the famous pots in Japan are originated from those in China.

I did not feel any conflicts in large-scaled Chinese Penjing. The world views for the environment and value are different, therefore people can feel the vigor full of taste and style from the magnificence of Chinese Penjing, and this is totally acceptable.

About Chinese Penjing and the world Penjing: The aesthetics view, concept, artistic conception, style, spirit and skill, etc. can be considered as the depiction model of Japanese bonsai. However, the spirit level of Japanese bonsai is the aesthetics and mental state of Zen, which is also an independent aesthetic consciousness. Japanese culture is symbolized by vanity and silence, and the bonsai is its visualized artwork. Therefore, attention paid to bonsai corresponds to attention to Japanese culture, and it reaches a consensus with Japanese aesthetic consciousness. I think the world bonsai is founded on the basis of Japanese bonsai and Japanese culture. The difference between the world bonsai and Chinese Penjing is the difference between Chinese and Japanese culture, and it shall be gone at the last. The importance is to show respect and understanding to the culture of other countries. The development of bonsai culture of the world is the coexistence of bonsais with characteristics of every country. Besides that, Japanese bonsai and Chinese Penjing have the significance of existence, and there is no need to uniform their names.

Just because of the differences of cultures, habits and traditions, the bonsais of the world are so interesting. I hope each other will have their own characteristic and I think it is necessary to understand the essence of bonsai / Penjing. Bonsais came originally from China, and as time flies, the cultures, traditions and habits of bonsais and Penjings have become totally different. We can say that Japanese bonsai art is formed from Penjing art by the development of Japanese aesthetic consciousness cultivation on the land of Japan. The future development of Penjing will start from one place, be cultivated in the culture and habits of that country, and then become the bonsai / Penjing with its own characteristics. I think that the existence of difference is the importance.

The live demonstration was for people to learn a limited skill in a short time, and it had a strong element of "Show". Japan also has held activities in this form by referring to Americans, but this kind of activities does not explain the existing state of the original bonsai. The original bonsai skill is not done by people, the bonsai is a natural object formed by the force of nature and the accumulation of time. People's skill is very limited. Due to the limited performance time, the immediate effect is contrary to the original creation of bonsai or Penjing. For all of the time, demonstrations are with problems. If the performance would contain the national characteristics and unique spirit, it would be very interesting. Hope in the future people can find a demonstration form which purposes at learning the nature of Chinese Penjing and pursues "the integration of nature and human", "artistic conception" and "spirit".

The development momentum of Chinese Penjing is vigorous. And, the most exciting is that the young Penjing creators and practitioners are enthusiastic and more people have focused on Penjing and made efforts to learn the skills. In China, there are so many excellent experts on Penjing, and it is really rare. Chinese Penjing is marching to the future with big strides, and it is a good fortune for all the Penjing practitioners!

「日本」須藤雨伯 景道家元二世

盆景の作品はすべて感心するものばかりで大いに勉強になった。嶺南盆景の盆景理念は多くを知ることが出来た。特に日本盆栽との違いとして、作品に関わる時間の長さを感じ感心致した。水石等については、中国賞石・奇石・山水賞石等、水石が区別され色々な価値観の中に成立していることは大切と感じる。我々は、中国賞石はこれから勉強させていただく。盆器については、我が国の名鉢は中国の名鉢であるから何も変わることはないと思う。今後は盆器の歴史や伝統を深く研究する必要があると思う。

大型盆栽は中国盆景として何の矛盾も感じない。環境と価値と求める世界観により当然異なるのであるから、中国盆栽の大きさは格調・風格があり元気を感じてよろしいと思う。

中国盆景と世界盆栽について：ゆえに日本文化の審美観・理念・風格・風趣・風韻・技術等は日本の盆栽の写しと考えてよろしいかと思う。しかし日本盆栽の精神性は禅の美学であり、心である。そして独自の美意識である。侘・寂は日本文化を象徴するもので、それを具現化した作品が盆栽である。だから盆栽の関心は日本文化への関心であり、日本美意識に共感していると思う。世界の盆栽は日本の盆栽・日本の文化が基本として成立していると思われる。世界の盆栽と中国盆景の違いは、日本文化と中国文化の違いであり究極は何も違わないと思う。大切なのは他国の文化を尊重し、理解する事そしてそれぞれの国の特徴ある盆栽が存在することが世界盆栽文化の発展と考える。私は日本盆栽があり、中国盆景があることが大切であり、呼称を共通する必要性はないと考える。

文化の違いや習慣・伝統の違いがあって世界の盆栽が楽しいのである。それぞれの特徴があることが望ましいと思う。しかし盆栽・盆景の本質を学ばなければならないと思う。盆栽と盆景は文化の違いや風土・習慣の違い、そして時間の流れに沿って中国よりもたらされた。そして盆景芸術は日本の土壌により日本人の美意識に培われ日本盆栽芸術と発展したわけである。源流はどこであれ、その国の文化や習慣に培われ、それぞれの国の特徴ある盆栽・盆景が誕生することが未来の盆景の発展と考える。違いがあることが大切と思う。

デモンストレーションは、短時間に学ぶ限りある技術でありショー的要素の強いものである。日本もアメリカのスタイルを参考に行ったもので、本来の盆栽のあり方を発表する場とはなりえる。本来の盆栽技術は、人が作るのではなく自然の力と時間の経過により自然造化として存在するもので、人の技術には限りがあり、従来のデモンストレーションのあり方は問題である。限りある発表であるから、即効的要素を要求され本来の盆景づくりや盆栽づくりには逆行する。国の特徴を生かし、独自のものが発表できれば楽しいと思う。今後は中国盆景の本質が学べるデモンストレーションのあり方を研究してほしいと思う。「天人合一」「意境」「気」これらを求める。

中国盆景の発展の勢いと力強さを感じた。そして最も嬉しいことは、若い作家や若い人が情熱をもって、また多くの関心を持ち盆景に精進していること。そして中国には、盆景に対して立派な盆栽の専門家が多数おられることを大変素晴らしい事と思っておる。中国盆景の未来が確実に成長している。これは盆栽人として大変うれしい事であった。

【丹麦】汤米·尼尔森

欧洲盆景协会成员国丹麦盆景协会会长

[Denmark] Tommy Nielsen
the President of Danish Bonsai Association

当我们置身于2013（古镇）中国盆景国家大展展会现场时，我不禁被盆景树木的尺寸和质量所震撼。做盆景展之间的比较不是一件容易的事，因为盆景与盆栽在某种意义上不是完全一样的。这次盆景展上盆景展品、古盆和赏石的摆放是很值得称赞的。展品上方的灯光给展品带来了更好的效果，在这一点上很值得其他盆景展学习。非常享受可以参观没有那么多拥挤人群的盆景展，这样就使我有机会可以慢慢地欣赏这些盆景展品。我希望我们可以有更多的时间去欣赏全部的展品，有些展品我还没有来及看，但是我相信所有盆景都是相当优秀的。

当我将盆景和盆栽进行比较时，我认为盆景和盆栽都有它们各自的独特之处。盆景虽然是采用更为自然的方式来制作树木，但是在我看来，盆景更像是通过欧洲方式制作树木。在欧洲，我们拥有特有的树木制作方式，我们的制作规则和日本人的制作规则并不相同。当我回家之后，我将我拍摄的照片拿出来和我很多的朋友们分享，他们现在正在思考怎样采用其他的方式来制作盆栽。盆景让人们考虑采用更大的空间和更为自然的树木来进行制作，这样也使得它们看起来都不一样。我们都可以从盆景艺术中学到很多，因为盆景艺术是盆栽艺术之父。我们在欣赏盆景艺术之余，也应采用适用于盆景艺术和盆栽艺术制作的精髓。

从丹麦人的角度来看，中国盆景的规模相比丹麦盆景而言大了许多。单独摆放时会令人印象深刻，但是当摆放在大型展会中并且周围的树木都非常高大时，我认为它们看上去就不是特别大了。最吸引我眼球的是一个有5棵桧柏的小品盆景，让我印象十分深刻的是，虽然展出的是一个小型的盆景树木，但是在看照片时却完全看不出大型盆景和微型盆景之间的区别。当将两个尺寸进行比较时，我的脑海中便涌现了一些问题。首先，大型盆景需要采用大量的枝条，这也需要很长时间来等待树木的成长。因此，需要数年时间才能制作一个漂亮的盆景。但是如果采用小型枝条的话，这就只是一个小问题了。在一棵小型树木上修剪小型枝条非常不容易，可能需要数年的时间，树木才能长成所需的形状和尺寸，所以制作大型树木的难点和制作小型树木的难点是一样的。

将中国的盆景技术和其他国家的盆景技术进行比较，我们可以相互学到很多东西。在我看来，中国的盆景树木给出了一个全景，因此和树干相比，小型枝条并不那么重要。中国的盆景制作技术更加注重整体效果，而日本的盆景制作技术则更加注重单一的枝条。

中国盆景艺术比较奇特的一点在于其不同的思维方式。令我印象深刻的是，我看到一盆中国盆景装饰着小型瀑布和流水，因此在欣赏盆景时仿佛置身于一个村庄或一个花园中。这种将小鱼放入池中并在盆景盆中配上小型房屋饰品的思维方式非常有创造力，和每天在欧洲所看到的非常不同，对我来说发展中国盆景艺术是非常鼓舞人心的。

关于中国盆景与其他国家盆景制作技术方面的区别，我很难回答这个问题，因为我并没有看过很多演示。但是我认为我们都在盆景树木上使用了金属线，并且全球各地的技术好像都差不多。我们之间的区别就是我们所生活的大自然，并且我们应该让盆景和盆栽在我们特有的自然环境中生长。

When we were at the Penjing exhibition I felt overwhelmed by the size and the quality of the trees. It is hard to compare the exhibition to other international exhibitions as Penjing is something else then bonsai, but the way the trees, pots and stones were displayed were something that was very fantastic, the light on the trees gave them a very nice look, this is something many other exhibitions could learn from. It was also nice to visit an exhibition where there was not many people so you had the time to look at the trees and admire them. What I could have wished for was that we had more time to admire the trees as i did not made to see all the trees but amazing it surely was…

When I compare bonsai to Penjing there are some things that comes to my mind, both bonsai and Penjing are fantastic in there own way. Penjing is though a more natural way of making trees, it seems to me like Penjing is looking more like the European way of making the trees. In Europe we have our own ways of making trees and we do not have the same rules as the Japanese people have. After I came home and have shown my pictures to many of my friends they are now thinking in an other way of making bonsai, Penjing makes people think more out of the box and make more naturalistic trees so they all not looks the same. We can all learn a lot about the Penjing art as the Penjing art is the father of the bonsai art, we shall admire the Penjing art and use what we feel is the best from both ways of making trees...

In a Danish point of view China Penjing are very large compared to Danish bonsai, this is impressive when you see them standing alone but when they are put in to a large fair and all the trees are big I think that they are becoming to common...What was having my attention the most was a display of 5 junipers in shohin size, what I think is so impressing about these trees is that the small tree when it is displayed and a photo is taken of it you can not see the difference between a large Penjing and a very small Penjing...When you compare these 2 sizes to each other there are some issues that is on my mind, first of all the large Penjing have large branches wish takes a long time to grow, it takes many years to make a nice Penjing but if you cut a small branch it is not a major problem...On a small tree there will be a big problem if you cut a small branch that it the wrong one and it can take years to grow a new one wish has the same shape and size, so the difficult in making big trees is just the same as making the small ones... When I look at the China Penjing techniques and other countries' I think we can learn a lot from each other, the Chinese way of making Penjing is in my opinion a way that gives the tree an overall look, the small branches is not as important as the main branches and the Chinese way is focusing more on the overall look then on the single branch as the Japanese do...

What is a fantastic thing about the Chinese Penjing art is the different way of thinking, I was impressed that I saw Penjing where there was small waterfalls and lakes it the pot and the tree was displayed as if you were standing and looking in to a village or a garden. The way of thinking where there was put small fish in to the lake and there were small houses in the pot was very creative and not something you see everyday in Europe...For me it was very inspiring to develop Chinese Penjing art.

What make us difference is the nature where we live and we should be making Penjing and bonsai as they will grow in our countries nature...

【日本】 森前诚二
日本盆栽大师
《WABI》杂志主编

[Japan] Seiji Morimae Japan bonsai master & the Chief Editor of *WABI* magazine

关于首届中国盆景国家大展的展品：盆景技术性造型成熟度方面的进步非常了不起！我已经感受到了盆景人的素养与艺术感的巨大进步，展品的种类和造型也非常广泛，颇为感动！在申洪良先生（他常与我一起交流和研究）的引见下，我参观了盆器展区。真的为宜兴陶瓷绵延数百年的艰苦历史以及保存展示的现状而欣慰！在水石方面，我领略了文房的世界，这种世界由不同于日本水石之美的审美意识所缔造。从整体来看，与日本相比，这3个展区像大河的洪流一样，具有日本盆景人失去的精神和热情。

关于盆景的审美理念，我觉得作品中已经包含了作者的感性和深深的诗情。在风格上，体量大，树形大，非常震撼！总之，展品中充满了细腻的"风格"，不禁让人想起二胡优美的声音，让人非常愉悦。在历史感方面，在中国这样的大国，不同地区具有各自的文化历史，因此，存在各自不同的盆景之美。在技术方面，我认为已经达到了日本的整体技术。会场的整体配置和设计以大型盆景为中心，结构非常合理。

的确，与日本盆景相比，中国的盆景非常大，具有压倒性优势——这与最终观赏盆景的场面大小成正比。日本国内，在室内狭小的空间内以挂轴等进行装饰，形成了人与盆景在心中对话的尺寸。拜访了很多中国盆景朋友的庭院以后才明白，中国人在观赏时，喜欢以比日本更大的庭院和展示设施为背景，我认为这种观赏方式是与大型盆景相呼应的。

各个国家所形成的历史和文化观各不相同。在日本，存在以"人与自然共生并融为一体"为基础的文化。另外，对于盆景来说，夸张的作品太强调人的存在，不受欢迎，在平淡中蕴涵很深的审美意识、好像体现了老庄（老子和庄子）思想的贤人型盆景被视为最佳。总之，以说明"人类该如何生存"的盆景为理想。在欧美，包括庭园在内，对于自然，以人管理的外观为美。例如：欧州著名的庭园都与外面的世界隔开。在日本，审美意识在于将自然中的山直接与庭院连接起来，而且，如果可能的话，人们喜欢将自己融入自然的气韵之中。在拥有数千年漫长历史的中国大陆，今后，反映各地历史性文化观的盆景作品将会日趋成熟起来。

樊顺利先生是中国著名的盆景大师，我有幸在这次的国家大展中观看了他创作作品。注视着他未完工的大型黑松，我感叹居然能在这么短的时间内制作出来与我所想象的完全不同的造型。这是这次国家大展对我印象最深刻的。

所谓的技术，去研究的话，人人都可以有某种程度的收获。但是，我很清楚的是，在感受素材中所蕴涵的树本身的姿态时，如果在自己的人生中不能与盆景进行深刻对话，将不能倾听到它的声音。现在，很多日本盆景人都在制作人们易理解、畅销的作品。就像人类具有个性一样，盆景也具有各自的特性。展示素材中隐藏的树的本性，这是盆景作家的最高境界，在这么短的时间内，樊先生展示给我的让我收获颇丰。

这确实是非常精彩的大型展览！在日本，受运营团体之间利害关系等因素的制约，这样大规模的展览无法举办。如果要说对下次展览的希望，我觉得很难同时将偌大的展览会场和大量的内容了然于心，因此，如果发放各展区的简介或者活动举办的简介，想看的内容就不会遗漏，岂不是更好吗？

在这次国家大展中，除了我所认识的人以外，也有很多志同道合的人，非常感动、欣慰。在中国盆景界，进一步地研究和发展将是注定的事情。我从15岁开始接触盆景，到2014年春天正好40年。为加深中日盆景友好交流，我愿意尽绵薄之力，希望以后能在这里发表作品。

About the exhibits in the first National Exhibition of Chinese Penjing: The progress in the technical modeling maturity of Penjing is pretty amazing! I have felt the huge development of Penjing artists' cultivation and artistic sense, and I am very touched by the varieties and modeling of exhibits! Thanks for Mr. Shen Hongliang's (whom I usually communicate and do research with) introduction, I have visited the exhibition stand of Penjing. I am so moved by the arduous history of Yixing ceramics of hundreds of years and gratified by the current situation of preserved exhibition! For scholar's rocks, I have experienced the world of scholars, which is built by the aesthetic consciousness different from the beauty of Japanese suiseki. Generally speaking, in the comparison with Japan, the three exhibition areas are like the river flood with spirit and enthusiasm which have been lost by Japanese bonsai artists.

About the aesthetic consciousness of Penjing, I think that the artwork has contained the emotion and the profound poetics of the artist. And I was shocked by the style of large body weight and magnificent tree shape! In a word, the exhibits are full of an exquisite "manner", which reminds people of the exquisite sound of Erh-Hu and is very pleasant for view. In the aspect of historical sense, in a large country as China, different areas have their own cultural history, which leads to the different beauty of Penjing. In the aspect of technique, I think it has reached the overall level of Japan. The overall layout and design of exhibition hall focuses on large Penjings, so the structure is very rational.

Indeed, comparing to Japanese bonsai, Chinese Penjing is very large with the overwhelming advantage – which is proportional to the final Penjing appreciation scene. In Japan, the small room decorated with hanging scrolls has formed the conversation space between people and Penjing. Only after visiting plenty of gardens of Chinese Penjing artists, I could understand that, in comparison with Japan, Chinese prefer larger gardens and exhibition facilities as the background for appreciation. I think this appreciation manner echoes with the large Penjings.

The historical and cultural views formed in each country are different. In Japan, the culture basing on "coexistence and integration of human and nature" exists. And for Penjings, the exaggerated artwork which emphasizes on the existence of human is not very welcomed. Penjings of virtue type expressing the thoughts of Lao – Zhuang (Laozi and Zhuangzi), which imply the deep aesthetic consciousness in calm, are considered to be the best. In a word, the Penjing explaining "how should people live" is ideal. In Europe and America, for the nature, including gardens, the beauty is expressed in the appearance managed by human. For example: the famous gardens in Europe are separated from the outside world. In Japan, the aesthetic consciousness is to directly connect the mountain in nature with the garden, and if possible, people like to blend themselves into the spirit of nature. In the mainland of China with a prolong history for thousands of years, Penjing artworks reflecting the historical and cultural views of each area will gradually mature from now on.

Mr. Fan Shunli is a famous Penjing master of China. I have the honor to appreciate his creation in the national exhibition. Looking at his unfinished large black pine, I am surprised that the modeling, totally different from my imagination, can be fabricated in such short time. This is the most impressive to me in the national exhibition.

During the research of the so-called techniques, people will gain in some aspects. However, I am very clear that when feeling the tree's own gesture contained in the material, its voice will not be heard if one cannot make deep conversation with the Penjing in his life. Nowadays, many Japanese Penjing practitioners are fabricating salable artworks which are easily understood by people. Like the personalities of human, Penjings would have their own characteristics. It is Penjing artists' highest state to show the nature of tree hidden in the material. In such a short time, I've learnt a lot from Mr. Fan's exhibition.

This is truly a very splendid large exhibition! In Japan, being restricted by the factors such as benefit-based relationships between the operating groups, the exhibition of this large scale cannot be held. Speaking of the suggestions for the next exhibition, I think it is hard to well appreciate such a large exhibition hall and numerous contents. Therefore, if there are introductions for each exhibition area or activities to be held, then we will not miss the interested parts, wouldn't it be better!

In this national exhibition, besides people I know, I have met many like-minded friends. I am very touched and gratified. In the field of Chinese Penjing, further researches and developments are definite. I have been in contact with Penjing since I was 15, and it has been 40 years till next spring. I would like to do my bit of effort for deepening the friendly Penjing exchanges between China and Japan, and I hope I will present my works here in the future.

「日本」森前誠二 日本盆栽作家 『WABI』雑誌主筆

初回中国盆景国家大展の出品について、盆景の技術的な造形完成度は、その進歩が見事である。盆景家の皆様の培養と芸術的感覚の素晴らしい向上を感じた。出品作品の種類と樹形も幅広く、感動した。盆器部門は、私と深い研究交流を持つ申洪良先生の案内で拝見した。宜興陶家の数百年に亘る苦心の歴史と保存展示される様子に感動した。水石につきましては、日本の水石美と違う美意識によって完成されている文房清玩の世界を感じる勉強になった。全ての点で、日本に比べて3部門共、大河の流れを見るように、日本盆景家が失ってしまった精神と情熱が存在していると思った。

盆景の審美理念については、作品の中に作者の感性と深い詩情が込められていると感じる。風格については、圧倒的な量感と大型の樹形に感心した。併せて繊細な「風を感じる」ような、まるで二胡の美しい音色を思わせる作品が展示されていることに喜びを感じた。歴史感は、中国という国の大きさが様々な地域による文化的歴史を持ち、それによって創出された盆景美があることを知った。技術的な面は、既に日本の整姿技術と変わらぬ域に達していると思う。会場全体の配置や設計は、大型盆景を中心に、見事に構成されていたと思う。

確かに中国の盆景は日本から比べれば大変大きく、圧倒されるが、これは盆景を最終的に鑑賞する場面の大きさと比例するものである。日本の場合は室内の狭い空間で掛軸などと飾り、人と盆景が心の中で対話する為の寸法になっている。多くの中国の友人盆景家の庭園を訪れてわかることであるが、日本より広大な庭と展示設備によって構成される観賞方式は、この大型盆景が相応しいと思う。

各国によって築きあげられた歴史と文化観は違う。日本の場合は「自然と自分が共生して一体となる」ことを基本とした文化がある。盆景でいえば、誇張された作品は人の存在が強すぎて嫌味があり、目立たぬ中に深い美意識を潜ませる、まるで老荘思想を体現した高士のような盆景を最上とする。つまり、「人はどのように生きるべきか」の答えを持つ盆景を理想とする。欧米は庭園も含めて、自然は人間の管理した姿に導く事を良しとする。例えれば、欧州の見事な庭園は外側の世界を区切って作られている。日本はまるで自然の野山の連携がそのまま庭へと続く美意識なのである。そしてできれば、その自然の気韻の中に自分を溶け込ませたい。数千年の深い歴史を有する中国大陸に於いては、今後各地に於いてその土地の持つ歴史的文化観を反映した盆景作品が熟成されていくことを望む。

中国の有名な盆景家である樊順利先生の創作を今回の国家大展で拝見した。未完の黒松の大木をじっと見つめ、私が思い描いた完成予想と変わらぬ姿を短時間で見事に作出された。この出来事は、今回の国家大展で私が一番深く感動したことである。

技術というものは、研究すればある程度は誰もが手に入れる事ができる。しかし、素材の中に潜むその樹が物語る本来の姿を感じ取ることは、如何に自分の人生に於いて盆景と深く対話しなければ、その声が聴こえないかは私もよく知っている。現在の日本盆景家の多くは、誰もがわかり易く、売り易い姿に作りがちである。人に個性があるように盆景にも全て違う個性がある。素材の中に隠れたその樹の本性を見い出す事が、盆景家としての最高の仕事であり、その点でも僅かな時間で樊先生が私の前で見せてくれた創作は見事の一言である。

今度の展覧会は素晴らしい大展である。日本では運営団体同士の利害関係などで、これ程大規模な展覧は行なわれたことがない。次への希望といえば、広大な展覧会場と多くの内容を一度に把握することが難しく、部門別の案内書や行事進行の案内書があれば、見たいものを見逃すことなく、更に楽しめると思う。

この国家大展で私が知る先生方の他に、多くの心を同じくする人達がいることに一番感動と喜びを感じた。中国盆景界は更なる研究と発展を続けるであろう。私も15歳より来春で40年の盆景歴を迎える。中日の盆景による友好と交流を少しでも出来るように、今後もこの誌上で研究発表をしたいと願っている。

【西班牙】 安东尼奥·帕利亚斯
《当代盆栽》杂志执行主编

[Spain] Antonio Payeras
Editor and Director for *Bonsái ACTUAL* Magazine

首届中国盆景国家大展给我留下深刻印象。这里展出的盆景作品是我见过尺寸最大的盆景，甚至与欧洲盆栽大会的合作中，我都没有见过如此大尺寸的树。这里有更多超级优良品质的树，好于在西方国家举办的任何其他的盆景展。本届中国盆景国家大展宏伟壮观。在这个展览上看到的盆景使我彻底改变了之前对中国树形的想象。

在西方国家，我们遵循日本的美学规则，所以我们的观点受这一特殊的盆景文化的影响。我认为中国盆景与其他国家的盆景最大的不同是分枝密度。此次参观访问中，我看到的树与西方国家和日本同样品质和成熟期的树相比，这里的树分枝密度较为稀疏。另外一点不同是盆的深度。同样规格的树，我们的树通常被栽种在比较浅的陶盆里。

正是这个差异给我们留下了深刻的印象。因为我们的盆景和日本盆栽，最大的尺寸大约是110cm。我想这是由于日本盆栽被陈列在壁龛里，所以壁龛限制了盆景的尺寸。在中国，盆景被摆放在园林里，所以尺寸不受限制。而且如果在宽阔的园子里摆放小盆景的话，不会引起人们的注意。我发现大尺寸盆景在中国显得更自然，比一棵独立的树栽种在此环境之外的地方更自然。我分析正是因为中国的园子很大，而且这个国家也很大，衬托出大尺寸盆景更自然。

对于大盆景的其他主要方面的顾虑是其可携带性。对于我们来说，一盆盆景必须是轻便、方便携带的，但是在中国，大的盆景树不需要那么做。因为这些树陈列的地点一般都在大的室外花园而且不经常移动。

通过在中国盆景国家大展上看到的现场制作表演，我认为中国的盆景创作技艺方面与我国的盆景创作技艺上没有太大差别。也许只是两国的盆景大师们使用的工具不同而已，因为两国树的尺寸不同，但制作过程是很相似的。其次，从中国朋友那里我学到了许多关于如何进行树的养护方面的知识，因为我们有着不同的土壤类型和不同的气候。

我们也组织了很多像这样的文化交流活动，我认为我们的盆景的表现手法将有所提高，因为我们在审美观念上引进了新的理念。在印度尼西亚我看到中国盆景理念与日本盆栽理念之间的联系，在那里，两种理念的混合被运用在他们的树上。

这趟中国之旅给我留下最深的印象是中国人民伟大的友善。我发现中国是一个现代并且有着良好社会关系的国家。对于一个美食爱好者来说，我发现这里有着品种丰富、样式多变的各种菜系。真是一件幸福快乐的事情。

I found the 1st National Penjing Exhibition was impressive. The area with trees displayed was the biggest that I never seen. Even cooperate to the European Conventions. Here has more trees with a superb quality than any other bonsai event held in the western countries.

The presentation also was magnificent. The Penjing that I saw in the exhibition make me completely change the image that I previously had about the Chinese trees.

In the western Countries, we follow the aesthetic rules from Japan, so our point of view is affected by this particular bonsai culture. So the biggest difference for me is the density of ramification. The trees that I saw in this trip have lighter branches than the trees that I saw in the western countries and Japan, in trees with the same quality and maturity. The other difference is the deep of the pots. Our trees are usually planted in shallower pots in trees with the same proportions.

This is the aspect that more impress to us. For us, as well as in Japan, the maximum size is around 110cm. That's because the bonsai in Japan are displayed in Tokonoma alcoves, so the tokonoma room gives the limit.

In China, the Penjing are displayed in gardens, so the size has no limit. And because in a large garden, a small bonsai will pass unnoticed, I found the big size of the Penjing in China more natural than if we analyze a standalone tree outside of its context. And the context is that the gardens are big, and the country is big too.

Concerning the other major aspect of large Penjing, is portability. For us a bonsai must be portable, but in China, large potted trees, need not necessarily be so, because its display place is always the same.

The creation techniques don't differ so much between our countries. Maybe Chinese Penjing masters use different tools because of the size of the trees, but the creation process is very similar.

I think that as much as we do this kind of cultural exchanges, all of our trees will be improved, because we introduce new aspects in our aesthetic baggage. I saw that relation between China and Japan view of bonsai in Indonesia, where a mixture of both ways are implemented in their trees. But for me, I learn a lot about how Chinese people do the maintenance of the trees, because we have different type of soil and different weather.

After this event, my biggest discovery has been the great kindness of the Chinese people. And I discover a modern and well connected country. And for me, a gastronomy lover, to discover the wide variety of dishes in Chinese cuisine it's been a great pleasure.

【日本】小林国雄
日本水石协会理事长
日本春花园 BONSAI 美术馆馆长

[Japan] Kunio Kobayashi the President of Nippon Suiseki Association and Curator of Shunkaen BONSAI Museum

这次的中国盆景国家大展非常优异，为各国的盆栽爱好者提供了交流的场地，宏大的会场可以尽情观赏。同时，我希望也能有壁龛装饰效果的展位。

对于中国的大型盆景，确实大的可以带给人强烈的“气”之感。但是，在日本的话，正如短诗、俳句（一种文体）中说的那样“形小相大”，可以在浓缩之中感受到广阔的空间和精神的深度。

盆景源于中国，当今的盆栽却是由日本人的感性和审美意识确立的。因此，希望盆景也可以表达出无法看到却无尽想象的闲寂、物哀等。

有实力的爱好者和企业家的支持是难能可贵的，应该借此大力推动盆栽的普及和发展。

中国对盆景的高度热情令人敬佩。今后世界的盆景主导权将会掌握在中国手中吧！

The China National Penjing Exhibition was very extraordinary and has provided places for the communication of bonsai lovers from different countries. You can fully enjoy yourself in the magnificent hall. At the same time, I do hope there are exhibition booths of tabernacle decorative effect.

For the large-scaled Chinese Penjing, it is really large and gives out the strong feeling of "vigor". However, in Japan, just as the saying in Haiku (a poetry form) "Small Shape Big World", we can feel the wide world and deep spirit in the concentration.

The Penjing origins from China, but the present-day bonsai is founded by the sensibility and aesthetic consciousness of Japanese. So I hope Penjings can also express the invisible but imaginable silence, sentimentality and so on.

The supports from substantial enthusiasts and entrepreneurs are rare and commendable and according to that the popularization and development of bonsai shall be vigorously promoted.

The high enthusiasm for Penjing in China is admirable. The future leadership of Penjing in the world may be held in the hands of China!

「日本」小林國雄　日本水石協会理事長　春花園 BONSAI 美術館館長

今度の中国盆景国家大展は非常に素晴らしかった。色々な国の盆栽を愛する人々の交流の場を与えていただいた。会場が広大なのでゆったりと観賞できた。床飾りを演出する余韻のある席飾りも欲しい。

中国の大型盆景はたしかに大きい方が強いエネルギー「気」をもらえる。日本の場合は「形小相大」と言う短詩や俳句のように凝縮された中に広がりと精神的な奥の深さを感じさせる。

盆景が生まれたのは中国であるが、日本人の感性と美意識によって今の盆栽の型が確立された。しかし、目に見えない物をも想像させる侘寂やもののあわれなども表現してほしい。

力のある愛好家や業者が協力していただけることは有り難いことである。大いに力を借り、盆栽の普及発展につなげるべきである。

盆栽に対する情熱の高さには敬服させられた。これから世界の盆栽は中国が主導権を握るであろう。

【法国】
克里斯蒂安·弗内罗
法国《气韵盆栽》杂志出版人
[France] Christian Fournereau, Chief Publication for *Esprit Bonsaï* Magazine

2013中国盆景国家大展给我们的第一印象是，太棒了！同样，我们也对盆景树木的质量、规模和数量留下了非常深刻的印象。所有盆景树木的造型不仅非常有趣而且还具有极高的水平。展会上大概展出了300棵盆景树木，但是我们在欧洲举行展会时一般只展出约80~100棵盆景树木。空间的使用和设置强调了盆景树木的最佳状态，但也确保参观的人可以在舒适的条件下来进行盆景作品欣赏。这次展会所提供的各种各样的树木和风格也非常有趣，落叶树种盆景非常不同于我们国家所在纬度所种植的物种。

我们有幸观赏到了许多我们平常无法看到的景观和丛林式盆景，而且盆景盆看起来也非常独特。此外，我们还看到了许多赏石，尤其是那些具有中国特色的石头。我们在欧洲展会很少能看到古盆，因此它们也给我们留下了非常深刻的印象。

对于中国盆景的第一感觉就是盆景的尺寸，中国盆景确实太高了。欧洲的盆景树木相比之下小了很多，我们无法想象居然能在展会上看到枝条长度超过1.5m的盆景树木！

如果说有些盆景树木的样式与我们欧洲的盆景树木非常相似，风格却截然不同。枝条相比而言不仅长度增加了，而且也展示出一个更为自由和更具创造性的风格。有些盆景树木的形状与树木在自然界中的样子非常相似，剩下的就是作者真正的创造，并展现了创作者的个人观点。总之，它更多的是一种诠释而不是复制。举例而言，“风吹式”盆景造型风格同样在欧洲存在。但是我只在中国见过树干往一个方向生长而枝条完全在相反方向的盆景树木！看起来太壮观了。在我们国家，盆栽艺术家经常试图开发一种非常明显能解释个人对自然看法的风格。中国艺术家似乎常常喜欢将中国老智者的雕刻融入在盆景中，借此将人类融入到树木和自然关系中。

中国盆景的“大”对于我们欧洲人来说，太让我们赞叹了也太让人意想不到了。他们同样也传递着强度和高能量感觉。

中国与外国盆景的造型确实不同，因此技术方面也会有不同，尤其是落叶树种的盆景制作。某些使用的制作技术非常少见甚至很多人都不知道。因此如果国际盆景艺术能运用中国盆景技术，将会有很好的效果。

这么短的时间很难了解中国盆景制作技术，然而中国盆景与外国盆景的操作方式却是相同的，都需要从需要被造型的树木开始，只在细微之处存在差异：金属丝蟠扎、修剪等。

在中国有机会能参观到两个特别棒的盆景园给我们的中国之行增添了更多的乐趣，我们也因此对中国盆景艺术有了更好的了解。

我们想对所有参与这次重大活动的准备工作和组织工作的人们表示衷心的感谢。我们有幸去发现中国盆景艺术并结识中国盆景艺术家。作为美食爱好者，我们还要对提供美味食品的人以及他们的热情好客表示由衷的感谢。我们想不出此处展会还有任何需要改进之处，唯一的一点就是时间太短导致我们没时间观赏完展会的所有环节。

对我们来说，中国不仅是一个开放和热情好客的国家，而且还是一个既保留了传统又展示了其开放性和现代性的现代化国家。它仍然是一个蓬勃发展的国家，并一直在建设自己的未来。这次活动让我们希望能再次来到中国并深入探索我们所看到的事物。

【法国】米歇尔·卡尔比昂
法国《气韵盆栽》杂志主编
[France] Michèle Corbihan, Chief Redactor for *Esprit Bonsaï* Magazine

The first impression we had about that exhibition was awesome! We were very impressed by the quality, the size and the number of trees. All the trees were very interesting and of a very high level. Concerning the number of the trees exposed, there were 300, when we are used to count about 80 to 100 in our exhibitions in Europe. The use of space and setting not only highlighted the beauty of the trees at their best but also allowed the visitors to admire them in smoothing conditions. The variety of trees and styles offered by this exhibition was also very interesting. The deciduous trees are very different from the species we growing in our latitudes. There was more landscapes and forests than we are used to see, the pots looked often exclusive and there was also much more original stones, typically Chinese. Ancient pots are rarely presented in the European exhibitions therefore we were very impressed.

Trees in Europe are smaller, and it is not imaginable to see a tree presented in an exhibition with a branch over 1.50 m long! If some of the styles are closed to the ones we have in Europe, the way the trees are styled is, however, very different. The tree branches are longer in a free and creative style. Some of the tree shapes are quite similar to the one that naturally occurs in nature but others are real creations, a personal view of the one who works the tree. It's more an interpretation than a reproduction.

The style "blown by the wind" for example exists in Europe too, but it's only in China that I have seen a tree trunk going in one direction, and the branches in the opposite direction! It is very spectacular.

The "large" Penjing in China for us are they are spectacular and unexpected for our European eyes. They also communicate a feeling of strength and high energy.

The differences between China and other countries is that the shaping of trees is really different therefore the techniques must be different too, especially for the hardwood species. Some of the techniques used are not commonly known of even unknown. It would be therefore fruitful for the international art of Penjing to share these Chinese techniques 。

It was too quick for us to have a clear vision of the China Penjing techniques. However the way of operating is the same, starting with a natural tree that will be shaped. The difference lays in the details: wiring, pinching, and pruning.

It was a great chance and a great pleasure too, to discover these two gardens and therefore to get a better view of Penjing art in China.

We would like express our warm thank to all those who took part in the preparation and organization of that great event and for their warm welcome. We were given a unique chance to discover the art of Penjing in China and to meet various artists. Lovers of fine food, we have appreciated the high quality of meals and the warm hospitality. We can't think of anything to be improved other than the time which was a bit short to appreciate all the richness of this event.

China appeared to us not only as an open and welcoming country but also as a modern country which has preserved traditions while opening itself to modernity. It's a flourishing country still building its future. This event make us want to come back and discover more about what we have seen.

【日本】镰田照士
《近代盆栽》月刊主编

【Japan】Terushi Kamada the Chief Editor of the monthly magazine *KINBON*

首先，展览的规模让我感到惊讶。其次，展出的一盆一盆精心创作的盆景让我感到震惊。

在现场制作表演中，作家的技术非常了得。与日本的盆栽作家相比，有过之而无不及，确实让我讶异。

特别让我感动的是，与日本盆栽又不相同的，只有中国才有的盆景的精彩。通过长时间创作而成的枝头的妙趣、在树根部摆放点缀性公仔以使作品风格和作者情怀更易于理解。这是我看到的与日本盆栽的不同之处。

我听说，在中国即使是一个出枝也要经过反复地修剪，其中的一个枝的创作也需要花费数十年的时间。在日本，使用金属丝进行枝条的短期创作，用时间、年月来赋予它苍古感和时代感。从这些技法和表现手法来看，中国盆景和日本盆栽也存在着根本的差异吧。

但是，这并不能说孰优孰劣。

石头、点缀性公仔的摆放使景色易于理解、呈现出大树观，从这种表现手法来看，这是中国盆景独有的文化，还记得我当时就被感动了。

总之，作为日本文化发展的盆栽实现了另一种发展（应该指日本式的发展），而作为中国文化被传承至今的盆景，则让人获得更多的感动。对此再次表示感谢！

Firstly, I was surprised by the scale of exhibition. Then I was shocked by every exhibited elaborate Penjing.

In the live demonstration, the artist has wonderful skills. He is going further when comparing to the bonsai artists in Japan. That also surprised me.

I was especially touched by the splendidness, which can be found only in Chinese Penjing, and it is totally different from Japanese Bonsai. The delight on the branch which has been fabricated for long and the dolls decorated at the tree root for interspersion make it easier for the artwork style and artist's feelings to be understood. This is the difference I found in comparison with Japanese bonsai.

I've heard that even one branching-out will be repeatedly trimmed in China, and the creation of a single branch will take several decades. While in Japan, we use the metal wire for the short-term creation of branch and the time and age will give it the senses of antiquity and times. According to these skills and expression techniques, there also is a basic difference between Chinese Penjing and Japanese bonsai.

However, we cannot tell which one is better.

The layout of stones and ornamental dolls makes the scenery easier to be understood and expresses the view of a magnificent tree. This expression technique shows the unique culture of Chinese Penjing. I remember that I was touched then.

In a word, as the development of Japanese culture, the bonsai has achieved another kind of development (which means the development in Japanese way here), but the Penjing, which has been inherited as Chinese culture till today gives people more sensations. Again, I would like to express my thanks!

「日本」鎌田照士『月刊 近代盆栽』の編集長

まずは展示の規模に驚かされた。そして出展された盆景の一つ一つがいかに丹精されたものであるかに驚かされた。

デモンストレーションでの作家の技術もなかなかのもので、日本の盆栽作家に勝るとも劣らない技術にも驚かされた。

特に感動したのが、盆栽とはまた違った、中国ならではの盆景の素晴らしさである。時間をかけて作り出された枝の妙味や、足元に添景を置いてのわかりやすい作風とこだわりには、いわゆる日本の盆栽との違いも感じさせられた。

枝づくり一つにしても切り返しの繰り返しで、中にはひと枝を作るのに何十年という年月をかけられているという話しも聞いた。日本では針金を使って短期に枝を作り、そこから持ち込みの時間、年月をかけて古さ、時代感を付けていく。これら技術と表現法自体にも根本的な違いがあるようだ。

これはどちらが優劣をつけられるものではないということでもある。

石や置物を置いての景色のわかりやすさや、大木観の表現法という点でもそれは中国盆景独自の文化であり、私としては素直に感動を覚えた。

いずれにしても、日本の文化として発展した盆栽とは別の発展を遂げ、中国の文化として継承されてきた盆景に、改めて大きな感動を得ることが出来ましたことをここに感謝申し上げたい。

【瑞典】

玛利亚·阿尔博尔莉思·罗斯伯格
瑞典盆栽协会理事会成员

[Sweden] Maria Arborelius-Rosberg
the Board Member of
Swedish Bonsai Association

我认为这次的展览非常漂亮，令人印象深刻，每个展品之间都有足够大的空间，一点都不显得拥挤。展场的灯光效果也很不错，展品的造像很有现代感，让人看了之后心情愉悦。其他的展会相比之下则显得空间很小，展品的摆放随之显得比较拥挤了。中国给我的印象是奢华高雅。中国盆景的规格大的令人吃惊，但是比例得当，所以看上去还是很赏心悦目。在每个盆景展品下面都铺上了桌旗，这个小细节十分有创意，如果能当做纪念品售卖的话将是一个很好的主意。我很喜欢桌旗的颜色。

中国虽然有很多大的盆景，但是这里也有小的盆景。所有的盆景都修整得十分整洁，而且你可以看出为了盆景漂亮整洁的外观，人们付出了多少爱和心力。我非常惊讶地看到，盆景造型的设计非常现代，给了我很多灵感，可以用在将来做我自己的树上。中国盆景的美学兼具传统和现代，看上去很愉悦。

从我看来国家间的盆景制作传统并没有特别多的不同，但是我能从中国盆景的创作理念和制作技术中学到很多，能从中国盆景大师身上学到很多！像我这样的北欧人终于可以逃离日本盆景“严格”的制作要求了。我喜欢中国盆景看上去就是一棵自然的树，但是又比大自然中的树更好看！经过这次的中国行程之后，我对中国传统抱有很大的敬意，希望所有人都可以享受这次中国之行。

盆景制作表演这种形式很好，因为可以向人们展示盆景制作的过程。制作盆景过程很复杂，但是经过多人的共同协作，很快就可以看出初步修整过后的效果。制作表演所用的原素材也让人印象深刻。如果在制作表演过程中可以加入英文解说的话那就更好了。

中国东道主的热情好客、慷慨大方让我印象极为深刻。每件事都考虑得十分周到，让我觉得受到特殊的款待。我向我的瑞典朋友讲述中国之行，他们听到后十分羡慕。我非常荣幸这次能有机会被邀请来到中国。中国是一个神奇的国度，有那么多漂亮的景色供人欣赏。我一定会与我的家庭再来中国的。我愿意学习中国博大精深的艺术和文化。中国在各个方面的现代化进程，都让人印象深刻。

再次感谢！祝愿大家幸福快乐！

I think that the exhibition was very beautiful and impressive because of the big space every Penjing had and that there was no crowding. It was very well lighted and the style of the exhibition was very modern and pleasing. Other exhibitions are more crowded and the space is much smaller. The impression in China was that it was very lavish and tasteful. The size of the Penjing was surprising, but since the proportions very so good it was just beautiful. It is Very nice to see the detail of the silk runner under each tree. It would have been a good idea if this beautiful item was for sale as a nice souvenir. I really liked the colors.

Of course the size is large, but there were also smaller trees. They were all very tidy and you could see that a lot of work and loving attention was used for the nice look. The trees looked very healthy and well cared for. I was surprised to see the designs that were new to me and gave me a lot of new ideas in the way I will design my trees. The aesthetics were both modern and traditional, but all very pleasing.

I don't think that there is so much difference in the tradition from that I know, but I think I can learn more from the Chinese way of creation and methods. With these nice results we can learn a lot from the Chinese masters! Nice to get away from a lot of the "strict" Japanese rules that we in northern Europe sometimes find restricting. I like that the Penjing look like a real tree, but better! After this trip to

China I have very high regards of the Chinese tradition and think all can enjoy this.

The demonstration was nice because it showed how they work, and that I recognized like we work. It was a lot to do, but since there were a lot of people working together it was fascinating how fast the result was visible. The high quality and size of the specimens was impressive! Maybe it would have been nice with an English explanation of how the design and work was planned before and during the demonstration.

I am very impressed of the fantastic hospitality and generosity of our Chinese hosts. I think everything was thought of and I felt like a special person. All my friends in Sweden are really jealous when I have told them of my wonderful experience! I was so happy to have been so lucky to have been chosen for this lovely event! China is a country that is wonderful to visit and has so much beauty for us to see, I will surely come back with my family again. I like to learn more of your fantastic culture and arts. Also of the modern progress in all aspects is very impressive.

Thank you again for this experience, I wish you all good in life and happiness!

【匈牙利】 阿提拉·鲍曼 EBA

欧洲盆景协会成员国匈牙利盆景协会副会长

[Hungary] Attila Baumann
the Vice President
of Hungarian Bonsai
Association

2013（古镇）中国盆景国家大展给我最大的感触就是中国盆景展览的规模以及盆景展品的尺寸远远大于欧洲国家。古盆展是个非常好的主意，在欧洲，古盆展览并不多见。展场灯光的布置非常专业，非常适合拍照。展品之间宽阔的场地，给人们提供了足够的空间参观和拍照。中国盆景反映着大自然的风貌，不用刻板的如日本那样遵循严格的规则。然而欧洲的许多国家却遵循着日本严格的技法要求。我认为，这次展会给我们提供了一个非常好的机会可以欣赏到中国真正高水平的盆景作品，从而体会到中国的盆景文化。

中国盆景与其他国家最大的差别在于盆景尺寸方面，在其他国家，规格小于100cm被称作盆栽，大型的盆景（至少在匈牙利）被称作"景观"树。在展台设计方面，许多国家不为每个展品单独放置一个展台，而是用一个或者几个很长的展台，而且盆景之间会有水石和添景（草本植物）相呼应，还有铜质的配饰、字画等。

在2013（古镇）中国盆景国家大展之前，我从来没有面对面的见过如此大的盆景树，非常震撼。盆景树的成熟度显示出展品的高水平还有多年的精心养护。我同样了解到盆景树的尺寸也反映着中国正在稳步发展的经济，这对于我来说还是一个比较陌生的概念。中国盆景的尺寸可以反映出更加平衡的枝条结构，叶子的尺寸也被更好地合理安排，所以比起小型的盆景，可以在细节上更好地反映大自然。

关于盆景技法方面，在所有国家中，最常用的方法就是蟠扎，出于我自己的理解，中国盆景不经常使用金属线，而是单纯地通过修剪枝条来达到造型的目的。我认为这种方法需要对各种树种精深的知识，如它们的生长习性和修剪后的变化。最后的盆景成型效果看上去非常自然，不像日本那样一味地遵循严格的方法。不一样的文化角度可以寻找到一个共同点促进盆景未来的发展，但是我们应该尊重各国独特的文化并且尝试去理解这些文化。

展场上的制作表演非常有趣，但是有一点我不太明白，我以为制作表演中不会使用金属线。用于制作表演用的坯材已经被修剪过，在我们国家使用在自然环境下采集的盆景坯材直接用于制作表演。所以我看到这之间的巨大差别，在匈牙利我们使用未经过任何修剪的原材料，而不是部分成型的树木（枝条构造已部分成型）。我想要知道你们盆景制作表演的目的是什么，重点是哪方面？盆景制作表演者是谁？使用的是何种技术？用于表演的树种又是什么？你们是想要现场的展示效果（枝条看上去都经过完整的造型）还是为了将来的效果（造型只是一个起始阶段，之后叶片还会继续生长，一些枝条也会被嫁接）？

总体来说本次活动的主办方非常优秀和专业。非常感谢付出辛苦工作的工作人员们。每天都有安排的满满的有意思的行程，所有的问题都能得到很满意的答复，在中国停留期间，我们得到了大力支持，我已经开始想念每个参观过的地方和那里的人了。参观过的大部分地区对于欧洲国家来说还是很陌生的，但是这些地方代表了高水平的盆景文化，应该加大力度进行推广。

来到中国之前，我对中国已经有了一些了解，我知道在最近20年，中国的经济开始腾飞。我非常喜欢中国，在中国游览到的景色也令我印象极其深刻。我曾经3次来到中国，但是这次除了北京我还去到了其他城市，让我从更多的角度来了解中国。中国的美味佳肴简直是太多了，还有许多我从未见过的蔬菜，简直是太神奇了。中国的城市规划非常惊人，在每条街上我们都能看到许多宏伟对称的建筑，每时每刻都有新的高楼拔地而起。我确信，未来的中国不论是在经济领域还是盆景文化方面都会扮演一个越来越重要的角色。

The size of the exhibitions is much bigger, than in European countries, and also the size of trees is bigger. I think, that displaying ancient pots is a very good idea, it is not usual in Europe. The lighting of the exhibition was solved professional, it was ideal for taking pictures. Between the trees there were enough spaces for walking and taking photo. China Penjing mirrors the nature, and not so strict rules are applied as in Japan for example. Many countries however following the Japanese strictly defined rules in Europe. I think, this exhibition was a very good opportunity to see real high quality China Penjing, and get feeling about the Chinese Penjing culture.

The most important difference between China Penjing and other countries' Bonsai is the size, in the other countries are trees below 100 centimeters called bonsai, the bigger ones are called (at least in Hungary) "shape" trees. In regard of display stands many countries do not use own display stands for each tree, but one or more long display stands, and the trees are presented nearly always with corresponding suisekis or accent plants(kusamono, shitakusa), bronze figures, paintings together, so not alone.

What I have thought about China Penjing is that I have never seen before such big trees as Penjing face to face, and the trees were very impressive. The maturity of the trees reflects a high quality work and several year maintenance and care. I have learned in China that the size of the trees reflects the growing business, which is not known in Europe as a reason of the size.

This size can represent more balanced branch structure and leaf size is more appropriate, than on smaller trees, so the tree can represent the nature in more detailed way.

In all other countries the most often used technique is the wiring, as I understood in China that it is not usual to shape the trees with wire, but only via cutting, pruning the trees. I think this method needs a deep knowledge of the different species, their grow habit and their reaction on pruning. The final results look very natural, but not a strict application of the rules as in Japan.

The different cultural perspectives help to find a common way for the development of the Penjing's future, but we should respect the unique perspective of the countries, and understand that perspective.

The demonstration during the exhibition was interesting but strange for me, because I thought that wire will not be used. The selected material was already well trained, in our country we are using yamadories (collected trees from nature) for such demonstrations.

So I see here a big difference, that in Hungary we are using raw material instead of a partly developed tree (means the branch structure is already created of a partly developed tree). We would like to understand the goal of your demonstration, what is in your demonstration in focus? The people, who are demonstrating, or the techniques, they are using, or the tree? Are you shaping the tree in such kind of demonstrations for now or for the future? In my opinion there is a big difference, to shape a tree for now (to look perfect with the actual branches) or for the future (the shape is only in a start phase and in the future the counter of the foliage will grow, and some branches have to be developed or grafted.)

In general the organization was very good, and professional, thank you again for Abby, Spencer and for all, who contributed to this event.

We had every day full with very interesting programs, all our questions were always answered, we have got the best support during this stay, and I already miss them and the places we visited. Most of the places are unknown for Europe, however these represent really a very high value in the bonsai or Penjing culture for the bonsai enthusiasts in the world, which should be promoted wider.

Anyone who is informed has some knowledge or impressions of China. We are aware of China's economic growth in the last two decades. I have found that China and the visited places have been quite pleasant. I have been third times to China now, but now I had the possibility now to see other cities than Beijing, and this gave me more aspect to think about. Another thing that I noticed is that China has many different food and dishes with never seen vegetables. This was great. The town planning in China is stunning, we saw powerful and symmetrical buildings on every street, and new high buildings are emerging by the dozens in all directions.

I am sure that China will play a major role in the coming decades as well as in economy as well as in the bonsai culture.

【越南】 阮氏皇

越南盆景协会主席

[Vietnam] Nguyen Thi Hoang the President of Vietnam Bonsai Association

这次2013中国盆景国家大展令人印象深刻，它不仅举办的非常成功而且组织有序。几乎所有展品都十分精彩并具有创造性。此外，展会还展示了盆景大师们的努力和创新精神。盆景展品在盆景盆、树木和石头之间都有良好的结合，并创造了独特的统一性。中国盆景展示了所有生物（树木）和非生命物体（石头）的伟大精神，形成了一幅美妙的自然画卷。

中国盆景非常注重小细节，因此创造出令人印象深刻的造型。中国盆景具有不同的风格，这也是传统美和现代美的结合。中国是盆景的发源地，有着悠久的盆景种植史。通过这次展会，我看到盆景在中国很受欢迎。因此，盆景的展出方式使得那些盆景的制作者们都感到非常骄傲，因为他们自己和他们的作品都得到一致好评。此外，此次盆景的展出方式也使得所有展出的盆景更加漂亮。

中国盆景的“大”代表了中国精神，它和中国的大自然文化一样非常壮丽。然而，如果想制作一个高水平的大型盆景，不仅需要一个能展示它的场地，还要艰辛地寻找到能制作大型盆景的大型树木和大石块，这些挑战也让观众更尊重这些展品。

我认为中国盆景的美学理念是“看起来越自然越好”。所有此次展出的盆景作品都看起来简单又美观。很难将我们的作品制作得如此优雅、简单并自然。它需要盆景大师付出巨大的努力和辛勤的工作。总而言之，这些盆景看上去虽然简单，但是相比复杂的作品而言，却能给人留下更为深刻的印象。每个国家都有不同的文化和风格，但因盆景是微缩的自然，所以尽管存在诸多差异，盆景必须看起来越自然越好。今天，许多展会和盆景会议将全球各地的盆景爱好者相约在一起。我们有机会讨论和交流经验，我相信我们会就盆景的审美标准达成一个具体的标准。

因为盆景起源于中国，所以中国盆景技术水平十分高超。中国和越南之间存在着许多相似的技术。盆景大师们对已成型的盆景进行改作的过程给我留下了深刻的印象。经过他们的制作之后，盆景变得更加好看了。因为古代历史的缘故，越南文化与中国文化有着较强的联系。因此，盆栽和盆景产品在审美、内涵和思想方面都极为相似。

我想对2013中国盆景大展的组织者们表示我由衷的感谢，我相信你们肯定是非常明智且热心的盆栽和盆景爱好者。这次活动非常有创意并且也组织有序。我真的非常感激你们，谢谢你们所做的一切。

中国拥有丰富的树木、植物和岩石资源，这些都可作为制作盆景的主要材料。此外，中国拥有悠久的盆景历史，我认为它也会拥有全球最优秀的盆景作品。当然，这意味着中国有许多的盆景大师，并且他们中的很多人都是高水平并且具备娴熟的技能。

The 2013 China National Penjing Exhibition this time is so impressive. It is great and well-organized. Almost every specimens displayed are wonderful and creative. Each of them shows the creative spirit and the effort of Penjing Masters. The Penjing specimens have the good combination between pots, trees, and stone, to create a unique unity. Penjing in China display the great spirit of living creature (trees) which are live well with non-living objects (stone), create a fabulous picture of nature.

Penjing in China pays attention on even small details, so it creates impressive shape. Penjing in China is diverse in style, and it is the combination of traditional and modern beauty. China is where Penjing originated; therefore, with the long-history of growing Penjing, during the development, bonsai and Penjing technique in China is the good selection of new and old culture, and then, creates marvelous Penjing products. Through the exhibition, I can see that Penjing is appreciated in China. Thus, the way Penjing being displayed make the Penjing creators feel proud of their products because them and their products are respected. Besides, the way Penjing being exhibited this time stimulated the beauty of each products.

The large Penjing shows the Chinese spirit. It is as magnificent as the big nature and culture of China. However, to have a good large Penjing, it requires big place to exhibit. Moreover, it is not easy to find a good big tree and big stones to make it. Those challenges make a good and large Penjing receive more respect from the audience.

I feel that the aesthetic concept this time is “look as much like nature as possible”. In general, all the specimens being displayed this time look simple but wonderful. It is not easy to make the products so elegant and simple but so natural like those. It requires a lot of efforts and hard-working of the Penjing masters. Simple but so much impressive than being complicated.

Each country has different culture and style. However, in general, Penjing is nature which is minimized. Thus, despite of all difference, Penjing has to be as much natural as possible. Today, there are many exhibitions and Penjing conventions which connect Penjing lovers around the world. We have chances to discuss and exchange the experience; I believe that we will come up with a solid agreement on the common beauty standards of Penjing.

The Chinese Penjing technique is good, and at high level of course, because Penjing is originated from China. There are many similar techniques between China and Vietnam.I am impressed by the performance that the Penjing master changed the style of a Penjing which is already good. After their work, the Penjing in the performance became even much better. Vietnamese culture has the strong connection with Chinese culture because of ancient history. Thus, the aesthetic and the connotation, ideas in the Penjing products are similar.

I wang to send thanks from my heart to exhibition organizers. Those must be wise and warm- hearted people who love bonsai and Penjing so much. The event is celebrated so creative and well-organized. I really appreciate that. Thanks for everything.

China has a great source of trees,plants and rocks, which are the major material to make Penjing. Besides, with the long Penjing history, I believe that China own an amount of the most excellent bonsai and Penjing products in the world. Of course, it means that China Penjing masters are so many and many of them have high skills and at high levels.

【韩国】 成范永
世界盆栽友好联盟顾问
"思索之苑"苑主

[Korea] Sung Bum-young
Advisor of World Bonsai Friendship Federation
Owner of Spirited Garden

我1996年第一次接触中国盆景。时至今日，再次来到中国受邀参加2013（古镇）中国盆景国家大展感触颇深，感慨万千。无论是规模还是作品都有很大发展，规格也很高。古盆和观赏石，数量和价值都相当震撼。展馆宽敞明亮，非常好。如果今后展场布置（如灯光照明）再精致一些就锦上添花了，我相信定会达到世界顶级水平。

过去中国大陆的盆景与韩国、日本、中国台湾存在很大差异，如今发展速度非常快。我实实在在地感受到，中国国家大，从事盆景的人多，潜力好，在短时间内取得了快速发展。看到，中国正在迅速摆脱固有的传统盆景的迹象，正向世界共享的树型发展。而且能看到，中国的树种和作品也在增多，其规模之大是在世界任何一个国家都无法看到的景象。总之，中国盆景事业取得了长足发展。

最初看到中国盆景感到十分新奇和惊讶。后来发现中国盆景普遍都比较大，不过，这也是符合国家特色的，国家大，作品大，看着也蛮不错。随着人们看到越来越多的不同的盆景表现形式，人们的意识也会随之发生变化。毕竟制作大的盆景作品在各个方面都不太容易。由于中国地域辽阔，所以中国盆景有不同的表现手法，有海派、岭南派、扬派、川派、苏派等，可以看出每个地方树型的发展都有所不同，而且因创作者不同的个性作品也不同。以前中国盆景与韩国、日本的盆栽存在差异，例如中国岭南派盆景不用铝线矫正树型以及盆土的使用方法等。我曾受邀在扬派盆景园为50多位盆景艺术家讲课。当时，我应邀对中国盆景艺术做评价。我说："盆景艺术与经济的发展紧密相连，随着时间的推移，会朝着购买者所需求的方向发展。"过去由于交通不便，中国大陆地域辽阔，每个地方都会有自己固有的作品。而如今，由于交通和通讯的发达，会迅速向着世界共享的树型方向发展。但我还是希望中国各地能够好好保护各自独有的盆景特色。

我觉得中国的盆景艺术已经进入高速发展的轨道。如果今后持续加以发展，那么不久的将来定会迎来中国时代。

我由衷地感谢主办方的精心安排和热情接待，祝贺本次活动取得圆满成功。对于下届中国盆景国家大展来说，毋庸置疑，一定会比这次办得更好。中国在各领域都取得日新月异的发展，期望不久的将来我们会迎来中国盆景时代的到来。

I've first come into contact with Penjing since 1996. It was very impressive and felt me great to visit the 2013 (Guzhen) China National Penjing Exhibition! Both the size and the works had all great development and the scale was also very high. The ancient pots and viewing stones, quantity and quality are quite good. The pavilion is bright and spacious, very good. If the exhibits should be more delicate and have an exquisite beauty in the near future, I believe that China National Penjing Exhibition would reach the world top level.

Chinese Penjing was quite different from Korea Penjing, Japan Bonsai before, but now it develops quickly. I indeed felt that China is huge with good potentiality and has achieved rapid development as well as many people are engaged in Penjing. I can see that China is not only rapidly getting rid of the signs of traditional Bonsai and is developing towards the trees of the world sharing, but also Chinese varieties of trees and works are rising, the scale is not seen in any country in the world. Anyway, China Penjing has obtained great progress.

I felt that it's amazing to see China Penjing at the beginning. China Penjing is generally large and also in tune with the times as well as the country is huge, so the work is big which is seen also pretty well. People's consciousness will also change. The reason why I said like this is that it's not easy in every aspect to produce a big work. It can be seen that the development of the trees in each place are different, there's the school of Shanghai, the school of Lingnan, the Yang style, the style of Sichuan and the Su style, etc, so the works are different because the author's personality is different, moreover, in many ways, for example, to correct the trees without the Aluminium wire and the usage method of potted soil, which are all different from Korea and Japan. But, China has started to use the Aluminium wire to correct the trees since 7 or 8 years ago, hasn't it? A few years ago, I was invited by Yangzhou city to give a lesson to more than 50 Penjing artists in the style of Yang's garden. I was asked for doing the evaluation to China Penjing art by then. I said that the art of Penjing would be parallel with the economy, as time went by, it would develop towards the direction of buyers' requirement. Due to traffic inconveniences in the past, China has a vast territory and every place has its own inherent works. But now China Penjing is going to go rapidly to the direction of the trees of the world sharing to develop because of the traffic and communication developed. But I still hope that every place around China can keep themselves inherent trees well.

I think China Penjing art has already entered the track of rapid development. If China Penjing should develop continuously in the future, China Penjing should ushered greet Chinese Era in the near future.

The exhibition organizers prepared for this event with great enthusiasm and 2013 (Guzhen) China Penjing National Exhibition achieved great fully success. I'd love to express sincere congratulations to them. For the next event, without a doubt, they can do better than this.

China achieved a rapid development in various fields and I look forward to welcoming the arrival of the world's first country, the G1 China Era in the near future.

【马来西亚】 蔡国华
国际盆栽俱乐部会员
马来西亚盆景雅石协会会员

[Malaysia] Dato'Chua Kok Hwa Member of Bonsai Clubs International, Member of Malaysia Bonsai N Stone Association

我对中国鼎——2013（古镇）中国盆景国家大展的印象很好，无论是从精心的策划、有序的组织还是执行落实上。这是一场独一无二的国家级盆景展览，不仅有盆景和赏石展，还有古盆和中国字画。我认为这届汇集盆景、石头、古盆和字画的全方面立体展览相得益彰，整体展现了中国盆景艺术文化。我感觉中国盆景已达到了堪比日本盆景和中国台湾盆景的水平。然而，有些方面有待改善，例如：盆景的陈列，盆景与人物配饰的搭配，盆景与赏石、字画等一起的摆放。

在我看来，大多数的盆景展，尤其真柏和松树类盆景展，中国鼎大展也差不多的，与在其他国家的盆景就造型而言，除了一些当地特有的盆景树种，它们都具有创造性和自己的个人风格。就盆景的造型、创造力、风格、美学而言，本届获得中国鼎冠军的盆景就是一个杰出的中国盆景的好例子。大展的展览大厅、展品陈列和展台设计都非常优秀并且看得出是精心的考虑、周详的设计。

我认为大部分的世界盆景都不超过1.2m。原因可能是容易养护、便于运输、从室外到室内灵活搬运便于观赏等。印度尼西亚是另外一个举办“大型”盆景展览的国家。我觉得盆景的尺寸被当地市场的需求所驱动并取决于盆景迷们的喜好。

在盆景创作方面，中国岭南盆景技艺是独一无二的。运用“蓄枝截干”技术增加盆景造型的独特性和美感。枝条的处理是盆景创作的重要方面而且所谓盆景美学就在于韵律、动感和枝的平衡。我觉得文化差异、个人的洞察力和阅历影响盆景的创作。我所看到世界大部分地区的盆景都由此创作出不同形式和风格的盆景作品。盆景是一件艺术作品。盆景一定是独一无二、有创造力、能表达盆景艺术者的思想和创作者每一丝细微的思想意识。我希望有一天盆景将达到中国绘画、陶瓷艺术之水平。

在盆景制作表演上很难看到盆景技术和创作的细节。盆景大师们都是在很短的时间内制作盆景而且大部分时间他们不做什么解释，急于完成盆景表演。另外，盆景制作表演上的大部分盆景几乎都是已经成型的盆景。

我由衷地想对苏放会长说声“谢谢”，感谢他的同事及全体员工。感谢他们所做的每一件事情从“中国鼎”大展一直到2013“唐苑的世界盆景对话”西安年度论坛。他们尽最大努力每时每刻提供最周到、舒适的服务。再次感谢！

中国是个很大的国家，给我留下深刻印象。中国政府大力支持推动中国盆景事业的发展。中国盆景已经发生巨大的转变以及中国盆景在世界盆景中的地位被极大地提升。我相信中国盆景将继续演变属于自己的风格和特性，而且将成为引领世界盆景中的一位。

My impression on the China Ding was very well planned, organized and carried out. It was unique to see a Penjing exhibition not only exhibiting Penjing and stones but also exhibiting ancient pots and Chinese paintings. I assumed the exhibits, Penjing, stones, pots and paintings are interrelated; represent the Penjing art culture in China. Personally, I feel that China Penjing has reached a level that is comparable to Penjing in Japan and Taiwan. However, there are areas for improvement such as display of Penjing, and the display of Penjing together with stone, plants, figurines and painting, etc.

In my opinion, most of the Penjing exhibited, especially Chinese Junipers and Pines, in the China Ding are similar to Penjing in other countries in terms of form except for some Penjing (local Penjing species) which are creative and have their own style. The China Ding's Penjing Champion is a good example of outstanding China Penjing in terms of form, creativity, style, aesthetics, etc. The exhibition hall, displays of exhibits, and design of booths were excellent and well thought out.

I think in most parts of the world Penjing do not exceed 1.2m in height. The reasons could be easy maintenance, ease of transporting, flexibility in viewing either outdoor or indoor, etc. Indonesia is another country that has "large" Penjing exhibitions. I feel that size of Penjing is driven by the local market demand and depends on the liking of enthusiast.

The China Lingnan technique is unique in bonsai creation. The development of branches using this technique adds character and beauty to the Penjing. Developing branches are important aspects of Penjing creation and the aesthetics of Penjing lies in the beauty, rhythmic, dynamic, and balance of branches. I feel that cultural differences and ones' perspectives and experience affect Penjing creation. Penjing that we see in most parts of the world follow certain forms and styles. Penjing is a work of art. Penjing needs to be unique, creative, and convey meaning and nuances of Penjing artists. I wish one day Penjing will reach the level of Chinese paintings, pottery, etc.

It is very difficult to see in details of Penjing techniques and creation in Penjing demonstrations. Penjing masters have very short time to do up Penjing and most of the time they rush to finish Penjing with very little explanations. Moreover, most of the Penjing in demonstration are almost ready Penjing.

I would like to say "thank you" to Mr. Su Fang, his colleagues, staff members for their excellent organization, hospitality, warmth, and everything from the China Ding all the way to Xi'an Annual Forum of the 2013 "Dialogue to the World Penjing" They had tried to make every moment comfortable to us. Thank you again.

China is a big country. China government is very supportive of promoting Penjing in China. China Penjing has transformed and improved tremendously. I believe China Penjing will continue to evolve into its own style and character and it will become one of the Penjing leaders in the world.

【日本】山田登美男
日本盆栽作家协会会长
[Japan] Tomio Yamada
the Chief of Nippon
Bonsai Sakka Association

中国的会场有与国家大展相符的大规模和大规格。

关于盆景，我深感这几年进步之大，虽然有的上盆时间还比较短，但是这个问题是可以通过时间来解决的。中国的盆景种类似乎有点少。听说，近年来中国大量购入真柏，但是我却认为中国原本长期培养的盆景价值更高。中国的古典盆景是特别优秀的。中国盆景人重视以对自然的热爱来研究造型美和自然美，盆景的魅力在于树本身具有的风格和树格。盆栽的理想在于，小中体现的树的巨大，苍老之态令人感动。我认为关注大盆栽与小盆栽各自的美是十分重要的。日本虽然自古以来，理想的大小就有85cm以内、一到两个人可以搬动的定论，但是最近日本也有大型的了。

关于盆器，中国的古盆十分丰富，特别是盆上的图案更加多彩。日本要提高盆栽的品位，就应该重视中国的古盆。我希望今后会出现与中国古盆和谐相配的盆栽。

关于水石，色彩丰富的石头很多，可以感受到这是与日本自然石（黑色为主要基调）截然不同的文化。

关于会场的布置、展台的设计等，我认为日本和中国都必须要对一盆盆景的空间处理进行研究。

现场制作表演的人存在着很多问题。现在，是否是过于展现技巧了呢？短时间内对于一棵树的整形存在着勉强与困难，可以看到植物受到了伤害。技术展示环节值得研究。

我觉得，虽然不同的国家拥有不同的文化观，但是未来的盆栽是没有国界的。未来一定是盆栽像沉默不语地生长着的植物之神一样地存在着，世界融合发展。

中国是日本重要的邻居，协力促使盆栽文化越来越繁荣地发展是世界盆栽爱好者的共同期望。我们盆栽人谨记这点，奋发学习吧！

The exhibition hall of China is with the large scale and profile conforming to the national exhibition.

About Penjings: I deeply feel the great progress accomplished in recent years. Though some Penjings haven't been transplanted into pots for long, this problem will be solved by time. The variety of Chinese Penjing seems to be little. I've heard that China has purchased Juniper in bulk, but I think the Penjing which has been originally cultivated for a long time values higher. The classic Chinese Penjing is extraordinarily excellent. Chinese Penjing practitioners attach importance to researching modeling beauty and natural beauty by their love for nature. The charm of Penjing is in the style and spirit of the tree itself. The ideal of bonsai is to express the magnificence of tree in small and touch people by the gesture of vicissitudes. I think that the size of bonsai is very important for appreciating the beauty of its own. Though since the ancient times the ideal size of bonsai in Japan is within 85 cm and movable by one to two persons, there also are bonsais of large scale recently.

About pots: China has abundant ancient pots, and the patterns on the pots are especially splendid. Japan should pay attention to Chinese ancient pots to enhance the taste level of bonsai. I hope there will be bonsais which can harmoniously match with Chinese ancient pots in the future.

About scholar's rocks, there are plenty of colorful stones, and I can feel the totally different culture comparing to Japanese natural stones (with the main color of black).

About the layout of exhibition hall and design of exhibition station and so on, I think both Japan and China have to do some research on the space planning for each Penjing.

There are a lot of problems with performers of live demonstration. For now, are they over flaunting the skills? There are reluctance and difficulties in modeling a tree in a short time and we can see that the plants have been hurt. Further research shall be done for the step of skill display.

In my opinion, though different countries have different cultural views, there shall not be national boundaries for bonsai in the future. The future is like the god of plant as bonsai growing in silence, the world will develop integrally.

China is an important neighbor of Japan, and it is the common wish of all the bonsai lovers in the world that the two countries will cooperate and promote the bonsai culture for more and more prosperous development. We bonsai practitioners shall bear this in mind and study hard!

「日本」山田登美男　日本盆栽作家協会会長

中国の会場は、国家大展らしく規模も大きくスケールの大きさを感じた。

盆栽については、ここ数年の間に素晴らしく進歩した感じを受けた。鉢持ち年数が若いものがあるが、そのことは時が解決してくれるだろう。中国の古典盆景は大変に良かった。中国の盆栽は種類が少ないようだ。近年中国は、真柏を大量に購入されたと聞いているが、中国の本来の長い時間をかけた盆栽こそがより価値があると思っている。自然愛を感じる造形美や自然樹の美しい姿を研究することが大切で、本物が持つ風格・樹格が盆栽の魅力である。盆栽の理想は、小さいながらも巨木観や老木観の形態に人々が感動するのである。私は盆栽の大小については、それなりの「美」がそこに存在することが大切であると考える。日本では古来、理想の大きさを一人か二人で持てる物で'85 cm 以内という定説があるが、最近は日本でも大きいものがある。

盆器については、中国の古鉢は特に盆上の景を一層豊かにしている。日本でも盆栽の品格を高める為に中国の古鉢を大切に扱っている。今後、中国においても古鉢に調和した盆栽の発表が望まれる。

水石については、色彩豊かな石が多く、日本の自然石（黒基調が主）とはかなり違った文化を感じた。

会場の配置、ブース設計などは、日本も中国も一鉢の空間処理を大切に扱う研究が必要と考える。

デモンストレーションのやり方にいろいろと問題があるように感じられる。現在、あまりにも技巧に走り過ぎているのではないか？短時間内に一本の木を整形することへの無理と難題があり、植物を傷めているようにも見えることがある。この問題は見せる場面の研究が必要である。

国によって文化観があっても、未来の盆栽に国境はいらないと考えている。

盆栽は「黙して語らず」で生きた植物の神のような存在であり、世界を融合できる未来像でなければならないと考えている。

中国は、日本にとって大切な隣人であり、協力して盆栽文化をもっともっと発展させることは、世界の人々（愛好者）も望んでいる！私共、盆栽人は、それを忘れることなく勉強に励む。

【印度】苏杰沙

印度盆景大师

[India] Sujay Shah

India Penjing Master

2013（古镇）中国盆景国家大展很壮观，展览展品给我留下深刻的印象！这的展品更有意思，没有被严格的规定所束缚。

中国盆景与其他国家的盆景的不同之处是中国盆景大且非常漂亮。每一盆盆景都是当地植物群、文化和哲学的映射。这些盆景看上去像来自大自然且不带人工技艺的雕琢。它体现了中国盆景枝条展宕变化之美和树干强健的体魄，不像一幅非自然人工雕琢之作。

中国盆景毋庸置疑的是比其他国家盆景大。我认为这个原因很简单。有可用的土地和资金，而且中国盆景在中国社会里普遍深受人们的喜爱。而其他国家有空间上的约束。就印度而言，盆景更多是女人们的爱好，但女人们没有足够的资金实力。正如在其他国家一样，印度没有多少男人对盆景感兴趣。

通过大展上盆景的现场制作，中国的树是我见过的最好的树之一。

最后，我向本届大展的筹备委员会表示衷心的祝贺。我真切体会到来自组委会非常体贴、无微不至的照顾。中国朋友热情款待世界友人，我们相处得就像一个快乐的大家庭！

中国是非常漂亮的国家！中国人民非常友善，从文化角度讲，我们有点相似，因为我们是邻国而且有着古老悠久的联系！如果以后中国技艺非凡的艺术大师到其他国家参观访问、传播盆景艺术，我们将热烈欢迎并深感荣幸！中国是最好客的国家！让我们感觉就像在家里一样！感谢可爱的国家，感谢美丽的盆景！

2013 (Guzhen) China National Penjing Exhibition was magnificent and very impressive! It was also more interesting because it wasn't structured with rigid rules.

China Penjing is huge and very beautiful. Each Penjing was a reflection of local flora, culture and philosophy. These Penjing look like that they have been brought from nature and no projected by man-made skills. It has a rugged beauty not like an unnatural finished photo.

China Penjing is definitely larger than others. I think the reasons are simple. There is availability of land and availability of funds. There is also a lot of popularity as people like it. Other countries have space constraint. In India it is more of a ladies' hobby and sometimes ladies do not have enough financial strength. Not many men are interested as in other countries.

By visiting this exhibition, I think Chinese trees are one of the best I have seen. Moreover, the 2013 Annual Night of China Penjing was very good and very creative! We liked it very much. We felt there was a personal touch involved and that was what made it stand out!

In the end, I must say congratulations to the organizers! We were taken care of very tenderly. The organizers have treated the world friends as a family!

China is a beautiful country! Chinese people are very loving and we are culturally a little similar as we are neighboring countries and have old ties! We would also love if your very skilled artists visit other countries and spread this art! China was hospitality at its best! We felt like we were at home! Lovely country and lovely Penjing!

中国鼎——2013（古镇）中国盆景国家大展
展品选拔筹备工作回顾

China Ding-2013(Guzhen) China National Penjing Exhibition Exhibits Selection and Preparatory Work Review

文：鲍世骐 Author:Bao Shiqi

经过大家的不懈努力，中国盆景国家大展暨会员作品展于2013年9月29日下午在广东中山古镇隆重开幕，并取得了圆满成功。本次展会组织有序，接待热情，布展规格高，展品质量好，极大多数展品融艺术价值与经济价值为一体，充分体现了中国盆景国家大展的时代性与厚重感，是中国盆景史上前所未有的一次盛会。下面就一些具体的筹展工作做一个回顾与总结。

2013年3月，展品选拔小组去福建莆田挑选作品，左起为曾文安、刘文和、廖金华、陈伟、苏放、鲍世骐、林文镇

2013年3月，展品选拔小组在福建厦门，右起为柯成昆、陈文辉、陈国健、魏积泉、苏放、曾文安、鲍世骐、王礼宾

2013年3月中旬，苏放会长带领评选小组在浙江宁波，左起为徐昊、鲍世骐、苏放、黄敖训、马建中

2013年4月，中国盆景艺术家协会会长苏放（左三）、名誉会长鲍世骐（右三）、常务副会长曾安昌（右二）、副会长黎德坚（左二）在广东东莞为大展选拔参展作品

2013年5月，展品选拔小组在湖北武汉。右起为中盆协副会长关山、副会长李城、湖北省花木盆景协会副会长兼秘书长刘志斌、中盆协会长苏放、常务副会长曹志振、名誉会长鲍世骐、中国盆景艺术家协会会员刘桂球

2013年5月，中国盆景艺术家协会苏放、鲍世骐、王选民等专家、领导在浙江台州遴选展品，左起为刘荣森、王炘、孙筱良、陈文君、苏放、管银海、鲍世骐、周修机、王选民

苏放会长、鲍世骐会长、刘永洪会长在四川

展品筹备小组在湖北，右起为陈光华、曹志振、苏放、鲍世骐

苏放会长（右四）与名誉会长鲍世骐（左三）、常务副会长夏敬明（左四）、林志远先生（右三）等人在温州选拔作品

一、展品的征集过程

本次大展展品选拔分南北两块区域同时征集。广东省盆景协会的任务是选拔40件参加国家大展的作品，同时在广东省选拔100件中盆协会员展作品，这140件由广东省盆景协会负责。广东以外的其他地区则征集大展作品60件，会员展作品100件，由中国盆景艺术家协会秘书处负责联系资料。

协会方面则由苏会长和我负责前往广东以外的省市现场预选参选盆景作品，我们预选小组的任务并不是评选入选而是先将全国较优秀的盆景作品筛选出来，并为入选评审组存档备案。另一方面，协会秘书处负责会员自行报名参展的作品，将能够参展的作品资料整理归案，两项相加，广东以外选区共获得预选及报名的盆景作品226件。随后，协会组织专家在上海对广东以外的区域选出这226件预选作品进行定审，最终选出60件大展作品和100件会员展作品。广东省选区则开始筛选广东本地的展品。以下是定审工作的具体过程：

时间：2013年7月30日

地点：上海颖奕高尔夫皇冠假日酒店会议室

评审组组长：苏放

副组长：鲍世骐

评委：曾文安、王选民、王永康、徐昊、樊顺利

监委：徐雯、苏春子

评审过程：

1. 第一轮评选：评委通过打√打X的方式，从72个申请人的共256盆盆景（微型盆景组合2组）预选出184盆入围展品。

2. 第二轮评选：评委对185号至225号票数相同的展品进行复选，按票数选出16盆入围展品。

3. 第三轮评选：评委对入选的200盆展品进行百分制标准评分、排名，按分数选出前40盆入选国家大展的展品。

4. 第四轮评选：评委对排名41号至90号初选入围展品再次评分，按分数排名选出20盆入选国家大展的展品。

5. 前60盆进入国家大展的展品因故不能参展，则从140盆展品中按排名顺序依次替补。前60盆以后则按得分的高低依次选出100件参加会员展的作品。以上评选过程均由评委独立进行。并由每位评委将自己的评分表签字确认。

广东省盆景协会十分重视广东地区参展作品的挑选工作。首先由国家大展组委会和广东省盆景协会联合向广东省内各地、市、乡镇的盆景协会发出了国家大展的通告内容和如何选送展品的要求和方法。

至2013年7月30日前，广东省各地要求参展的人选向广东省盆景协会、各地盆景协会报送了相关的资料，先由各地区协会组织第一次挑选，然后将预选出的作品汇总后报省协会。广东省盆景协会再根据报送上来的汇总资料情况，省协会决定组成两个专家组到各地市去把关挑选展品。由中国盆景大师、中国盆景艺术家协会常务副会长、广东省盆景协会会长曾安昌和省盆协副秘书长何焯光带一个组到顺德、中山、阳江、湛江、韶关等地区对报名参展的作品实地挑选；由中国盆景艺术大师、中国盆景艺术家协副会长、广东省盆景协会常务副会长秘书长谢克英和广东省盆景协会专家组组长余镜图带一个组到广州、佛山、南海、乐从、东莞、汕头等地区对报名参展的作品实地精选。

两个专家组成员所到之处都虚心地听取当地协会领导和专

曾安昌会长率广东省盆协专家组在遂溪县盆协遴选参展作品

中国盆景艺术家协会副会长谢克英和广东省盆景协会专家组组长余镜图带队在乐从进行展品选拔工作。前排左一蔡有德、左二陈有浩、左三谢克英、左四余镜图、左六蔡汉明、左七蔡景初，后排左一吴利雄、左三王金荣、左七陈华勉、左八许耀声

广东省盆景协会会长曾安昌率专家组于湛江盆协基地遴选作品

中国盆景艺术家协会常务副会长曾安昌（右五）和广东省盆协副秘书长何焯光（右四）带领选拔小组在广东澄海选拔参展作品

家人员及作者本人的意见后，省协会专家组人员反复和他们沟通交流，取得较为一致的共识后才初步选定要参展的作品，挑选作品既考虑规格要求，更注重质量和代表性，原则上一个送展者送展的作品不要超过两盆。两个专家组不畏艰苦，不怕疲劳，连续作战，历时一个多月，驱车数千千米，从预选的数百盆作品中，初选出了参加国家大展的43盆，参加国家协会会员精品展105盆。在这基础上，于8月30日前省协会两个专家组一起专门召开了会议，对各专家组初选出的作品资料放在一起又一次进行认真地、一丝不苟地对照筛选，最后决定了广东地区参加国家大展的展品37盆（广东地区应选展品40盆）；参加协会会员精品展的作品91盆（广东地区应选作品100盆）。

至此，中国盆景国家大展暨会员展的参展作品基本尘埃落定。

二、签定协议、发放参展须知

为确保展品的落实，由协会与入选展品的作者签定参展协议，同时制定了详实的参展须知，一并发放到参展作品的作者手中，提醒作者作品入选后的养护及美化工作，确保每一件作品都能以最佳的精神面貌进入展会。

三、组成大展评审委员会

评比工作的好坏，也是大展成败的关键，因此，协会制定了选择评委的条件以及评委组成的原则：

1. 由大展组委会聘请业内资深专家组成评审委员会，评审委员会包括评委和监委，评委不少于7人，监委2~3人。

2. 评委和监委成员必须在盆景界中具备专业性和权威性，具有良好的职业操守。

3. 评委会设评委主任一名，副主任一名，负责召集评委及监委，主持开展评审工作。

4. 组委会在适当的时间向评委、监委人员发出聘书，评委、监委成员在收到聘书后至评审工作开展前，要求严格保密。

5. 在评比前，评委会须召开一次会议，明确评比要求、评比办法和注意事项等问题，统一认识，明确任务，再进行正式的评比工作。以下是经协会常务副会长会议及组委会讨论确定的评委会组成人员：

评委组主任：谢克英

评委组副主任：王选民

评委：徐昊、曾文安、陆志伟、罗传忠、徐闻

监委：樊顺利、魏绪珊

四、制定评奖工作原则

1. 遵循公开、公平、公正的原则。

2. 评委、监委的参展作品不得参加评比。

3. 坚持标准、认真负责、尽量减少失误；要对每一件作品的评审结果负责。

4. 评比时，评委必须单独对作品评审打分，不得相互商议，不得互通评比意见和结果。中途评委要离开现场的，必须将评分表交给监委，回来时再拿回评分表继续工作。

5. 评比过程由监委全程监督。

6. 评委打分完成后，要在每张评比表上签上自己的名字，以便查询。

7. 评比结果公布前，评委不得泄露评奖结果。

五、确定评比标准

1. 题名 5分

题名确切，寓意深远，是形式与内涵的高度概括。

2. 景 80分

善于运用盆景艺术创作原则，具有熟练巧妙的造型手法和优秀的栽培技术，作品生长健壮，成熟度高，构图优美，布局合理，而且具有丰富的中国文化内涵，达到形神兼备、小中见大、源于自然、高于自然的艺术效果。

3. 配盆 10分

配盆的形状、大小、质地、深浅、色泽与主题得体，使盆与景相得益彰，盆面的地形地貌处理得自然、美观。

4. 几架 5分

几架造型、大小、高矮、色彩与盆景配置协调，能充分体现盆景之美，使盆景达到最佳观赏效果。

5. 作品规格

高1.5m内（文人树最高可放到1.6m以内），盆长1.8m内，悬崖飘长1.5m内。

注：树的高度应从盆面起至顶端绿色（活的部分）位置计算，悬崖盆景飘长应从盆沿算起。

六、明确评比办法

1. 评比采取百分制综合打分的办法，评委根据组委会决定的评奖数额，定出了如下评选办法：参照评比标准，对每一件参展作品进行综合打分。为了使每个评委把握好评比的尺度，国家大展的参展作品评审时，对每盆作品的打分不低于65分；打90分以上的作品不超过9盆，打最高分的作品只能是1盆。按分数高低选出国家大展的9个奖项。中国鼎及8个大展奖分数重叠时，由评委进行复评，由监委主持，对重叠分的作品由评委举手表决，以多数为准。

2. 对会员展的参展作品的金奖在打分时控制在91~98分，银奖控制在81~88分。铜奖控制在71~78分，每个等级中间间隔2分，以防止重叠分过多而难于最后评定。如果计算出奖项的结果出现奖项末尾重叠分时，由评委进行复评，由监委主持，对重叠分的作品由评委举手表决，以多数为准。评委对每件参展作品打分不低于60分。

评分的计算方法是：评委评完后，将评分表交给监委，在监委监督下，由专业人员用电脑软件计分，对每件作品的评分，按预设程序去掉一个最高分和一个最低分，以剩下所得总分的平均分数为该作品的最后评定分数。然后按每盆作品的得分多少，从高分到低分按顺序排出中国盆景国家大展及会员展的奖项。评分结果出来后，由评委、监委到展场核对评审结果，核对无误后，评委、监委在评比结果总表上签字确认，然后评委会将评比结果交组委会公示。

由于组织工作细致、严密，使本次大展无论从展品质量、展会设计布置的档次、评比工作以及中国盆景年度晚宴、现场制作表演等活动，都取得了前所未有的成功。这里，我特别想说的是：感谢大家为本次大展贡献的智慧与辛勤劳动！感谢广大盆景人的努力付出！相信只要大家团结、努力，下一届的中国盆景国家大展一定会办得更好！

注：本文中涉及广东省盆景征集过程及图片部分由广东省盆景协会提供资料，涉及评比标准及评比办法由筹委会评比组提供资料。

谢克英会长和王金荣秘书长在澄海遴选展品

曾安昌会长率广东省盆协专家组赴廉江市盆协遴选参展作品

曾安昌会长及乐从盆景协会在澄海挑选参展作品

梁悦美教授，中国盆景艺术大师，国际首席盆栽大师，ABFF 亚太盆栽友谊联盟前理事长，中华盆栽艺术台湾总会名誉理事长，中国盆景艺术家协会名誉会长，台北市树石盆栽协会前理事长，美国西雅图太平洋大学及南区大学园艺设计、盆栽教授，国立台湾师范大学及中国文化大学园艺设计、盆栽教授。

An Interview with Professor Amy Liang, the International Bonsai Master

用爱编织我们的中国盆景艺术家协会

国际盆栽大师梁悦美教授访谈

受访人、供图：梁悦美 访谈人：CP
Interviewee / Photo Provider : Amy Liang Interviewer: CP

中国鼎国家展的颁奖晚宴上梁悦美教授荣获“中国盆景两岸文化交流终身名誉奖”

中国鼎国家展

这次中国盆景艺术家协会创办的“中国鼎国家展”，是有组织、有创意、经过慎重考虑与精心策划的标志性的创举，对中国盆景的前景有百尺竿头更进一步的作用，对中国盆景艺术家协会进入一个新的里程碑具有重要的意义，本人深感欣慰、可喜可贺。

2010 年我荣聘为中国盆景艺术家协会第五届理事会名誉会长，这次“中国鼎国家展”的颁奖晚宴上我荣获“中国盆景两岸文化交流终身名誉奖”，并

梁悦美教授在中国鼎国家展的展场上流连忘返

以中国盆景艺术家协会名誉会长的身份，颁发多项大奖，让我深深感觉到从1991年参加首次中国盆景国际会议起23年来在大陆出钱出力、教学、鼓励、推动盆栽的努力，得到了肯定和尊重，感到万分的欣慰与鼓舞。自己暗暗地在心中决定要终身无怨无悔，继续为中国的盆景奋斗，并将中国盆景文化推广到全世界。

以下是我对中国鼎国家展的感言与建议：

中国鼎国家展展场布局：

展览场的布局，是我见过世界上目前为止最令我赞赏、最高雅、最有文化水平的一场国家大展。

这次参展的盆景，每盆都有专属于自己的独立展台与背景。以“中国风”形象、中国独特精致雕刻图案作为图腾的展览台，使放置在展台上的盆景细致、流向、气势、风格、文化气息表达得淋漓尽致，令人感佩与震撼。

多年来我不断应邀至世界各国盆景协会进行专题演讲及示范表演，足迹踏遍全球21个国家，并分别以汉、英、日、台语授课，与学生充分交流。此外，曾受邀参加140场以上的国际盆景研展，担任盆景评审及剪彩活动，活跃于国际盆景艺术舞台。

1984年1月29日新年，本人在最著名的台北市市立美术馆举办为期10天的“梁悦美树石个展”。当时电视、电台、《新生报》及《中国时报》、《盆栽世界》杂志（第五期）均长篇大幅刊登：“梁悦美教授是中国台湾第一位将树石盆栽引进大学及进军美术馆，带进文化的殿堂，大幅提升树石艺术的地位，获得艺术界的肯定。她也是台湾首位创‘盆景单桩立体布置法’之人，她有五盆盆栽被台湾发行成五张台湾邮票。美术馆反应热烈，引起一股盆景热潮，并未受美术馆收门票影响，10天内涌进20万参观人潮。”

置身中国鼎展场，令我缅怀30年前在台北市美术馆的个展情景，不仅令我感到“于我心有戚戚焉”的感觉。盆景展至目前为止，全部盆景都是一排排整齐放置，没有上下起伏的排列方法。30年后的今天，终于有一场展览，是用“单桩独立布置法”展出了，我感动肺腑、潸然泪下，久久徘徊不能释怀，不舍离开。

中国鼎国家年度晚宴：

另外，这次展览给我印象最深的就是年度晚宴，这次的年度晚宴是有高水平的晚会，可圈可点，令所有的嘉宾赞美，尤其是国外嘉宾，他们事后和我谈起，都有口皆碑、赞赏不已。晚宴的气氛、排场尤其音乐及灯光颜色效果是我参加过所有的盆景晚宴中最好的。为获奖嘉宾特别设计登台的“T”型伸展台十

在西安“唐苑”举办的世界盆景论坛现场一角

梁悦美教授向世界盆景友好联盟前会长、国际盆栽俱乐部前会长索里塔 · 罗塞德夫妇介绍中国鼎国家展

两位中国盆景艺术家协会的名誉会长为中国鼎国家展感到欣慰（左：鲍世骐）

分新颖，有画龙点睛之美，使所有颁奖人及获奖的人显得分外荣耀与光彩。

模特儿们手捧盆景走秀更是令人瞩目。一个个打扮时尚长相靓丽的模特儿，把一盆盆小巧引人、不同造型的盆景托在纤纤手中，缓缓在展台上旋转展示。富有文化古艺术的盆景与新时尚潮流现代化的模特原本是格格不入的，但是经过灯光、音乐、色彩、笑脸、飘逸的长发、曼妙的脚步……声、光、色、影调和融汇一起，却是那样地别有一番风味，令人折服，别开生面。

颁奖环节虽然有多种奖项，但是各个奖项分类明确、有条不紊，全场感到很得体，令人有很好的回忆。

我对展览有一些建议与观点：

我建议：1. 展区将成品完美盆景与略显粗乱的半成品盆景，分不同的展区摆放，使展区不凌乱；2. 展览区标示要清晰明显，以便观赏盆景的人遵照指示行走，井然有序；3. 有专业老师负责导览全场解说，使观赏者了解如何管理与观赏；4. 在展场设置休息与喝茶区，分批轮替，使疲惫的嘉宾可轮流休息（这次展览场的茶区放置于后面偏僻处，很难被发现）。

趣怡园、真趣园、唐苑

中国鼎盆景国家展之后，国内外嘉宾团安排到趣怡园、真趣园、唐苑去参观。三个园子的主人都非常热情地接待，他们以树会友，深情款待，大家相谈甚欢，互相切磋盆艺，其乐融融，让每个人都感到很温馨，除了感恩还是感恩。

梁悦美教授和中国盆景艺术家协会常务副会长吴成发（左）共享

30 年前梁悦美教授在个人展中首次创立“盆景单桩立体布置法”

三处庭园，都有很宽敞的空间可充分发挥，非常可贵，也很羡慕。都有美好未来可展望。经一年又一年的努力与造景相信会有更加美好的愿景。

我的建议是，在广大的庭园土地上可以将素材盆景、半成品盆景、成品盆景分开，大型、中型、小型盆景、超大型盆景、庭院艺术树及古盆区等都要有条有序、分门别类，也可用上下左右、高低起伏来摆盆景，用三度透视及大小不同空间将盆景展示出来。庭园中指示路牌要分明，如果能将树种、树龄、习性及特殊管理方法或照顾方法注牌说明更好，如此可给观赏者更多了解与学习机会。

世界盆景论坛

世界盆景石文化协会和中国盆景艺术家协会在西安“唐苑”举办的世界盆景论坛是一个很好的企划。国际性盆景展览，虽然常常也有小型的国际盆景论坛，但这是首次较大型国际盆景论坛。许多世界盆景界真正知名人士汇聚到一起，报告每个国家盆景概况及进展，将全世界的盆景拉近，这种构想很有意思的。通过这个论坛，互相交流共同进步。演讲嘉宾用 PPT 图文并茂地讲解自己国家的盆景协会、盆景历程、盆景风格、盆景造型、盆景媒体等，让我们了解到世界各地的盆景发展进程，拉近全球盆景的文化。

此次论坛遗憾的是，还有一些在世界上真正知道盆景、有知名度、有影响力的盆景人由于时间关系没有来得及邀请到场，我希望下次的论坛活动，可以有更多人参加，论坛内容将更加精彩丰富，达到真正的目的。

为了“无言的诗，立体的画”的盆栽艺术，期待大家同心协力共勉！

30 年前梁悦美教授在台北市立美术馆举办了个人树石展

国内外嘉宾团展后参观，其乐融融

中国盆景国家大展观后感
Impressions of China National Penjing Exhibition

文：李正银、罗传忠 Authors: Li Zhengyin & Luo Chuanzhong

作者简介

李正银，中国盆景艺术家协会常务副会长，广西盆景艺术家协会会长，广西壮族自治区柳州市银阳房地产开发有限公司、广西照顺资产经营有限公司、柳州鑫盟投资担保有限公司、广西北海市银阳园艺有限公司法人代表。

纵观2013（古镇）中国盆景国家大展，我们认为，这是一次在中国盆景史上具有划时代意义的、代表中国盆景最高水平的顶级赛事。它之所以吸引人的眼球，不仅仅是形式新颖、富有创意，而且内容厚实——"中国鼎"的角逐以及100盆国家大展作品的入选，其本身就极具吸引力和挑战性。

国家大展值得肯定的做法有以下几个方面：

一、定位高、办展思路明确

中国是具有五千年历史的文明古国，也是盆景的发源地，随着中国国力的增长，人民生活水平的提高，中国盆景国家大展就自然而然地应运而生，这是历史的沉淀，时代的呼唤！"打造面向全球的中国盆景国家大展的标志性品牌，向全世界展示中国盆景国家大展的国家形象"，这一定位不仅具有划时代的意义，也代表广大中国盆景人的心声。正是由于这次展览定位高、办展思路明确，才使整个展览的设计策划，包括入围盆景和获奖盆景数量的限制、国家大展中国鼎首席大奖的设立、入选盆景地域范围的扩大以及场馆的布置等，都进行了有益的探索和改革，并且收到了预期的效果。

二、展前筹备工作充分

这次展览是历次国家级展览准备得最充分的一次。其中，做得最出色的是宣传工作，中国盆景艺术家协会利用《中国盆景赏石》这一平台，向全球盆景人发出了举办首届国家大展的通告，而且苏放会长在书中发表了多篇关于国家大展的高水平文章，对展览的有关事项作了具体的介绍，让广大会员及盆景爱好者充分了解参展的要求，提前做好参展的准备工作。此外，这次展览采取考察小组实地考察甄选、各省盆协推荐及作者自愿报名相结合的办法，作品入选地域范围扩展到所有落户于中国的顶级盆景，确保入围作品具有广泛的代表性，最终入围作品由国家大展入选评比委员会确定，一改以往参展作品随意性较大的办法，在一定程度上确保了入围作品的质量，而且还严格发出入选确认书，防止随意更换入选作品，这些都是值得推崇的好办法。

三、2013中国盆景之夜既别致又新颖

开幕式当晚的古镇中国盆景之夜，聚集了中外盆景界的精英人物，可谓高朋满座，群星灿烂。整台晚会气氛既隆重又热烈，简洁明亮的灯光舞台设计、轻快的音乐旋律、各

作者简介

罗传忠，中国盆景高级技师、广西盆景艺术大师、广西盆景艺术家协会专家委员会副主任。

国模特手捧盆景美妙的台步造型，给与会者带来了温馨浪漫的听觉和视觉享受。尤其是会长团队和中外盆景界重量级嘉宾的亮相，以及2013中国鼎、年度先生、年度城镇、年度协会等多项奖项的颁发仪式，将晚会气氛引向了高潮，使晚会真正办成凝聚力量、激励斗志、弘扬国粹的中国盆景年度盛典。

四、盆景质量上乘

这次共展出100盆国家大展盆景和200盆会员精品盆景，南北多种流派、多种风格的顶级盆景同时登台亮相，充分展示盆景发源地中国盆景丰富多彩的盆景艺术风采。总体感觉作品成熟度较高，基本没有出现以往盆景展览半成品甚至秃树充当精品的滥竽充数现象，整体作品质量较均衡，甚至在会员展里的个别作品比某些国家大展里的作品还要精到。独树一帜的岭南盆景依然抢眼，枝法、层次和布局更趋合理成熟；攀扎类盆景在保持原有中国盆景注重意境的基础上，善于吸收借鉴外来盆景好的理念和技法。总之，近年来的中国盆景在坚持不断创新，向源于自然、高于自然，更诗情、更写意的方向发展。

当然，由于国家大展是首次举办，很多东西都在探索和实践中，还有不少需克服和改进的地方。

在展品质量方面：

100盆入选的作品中，还有个别质量一般的作品，需要主办方在严把质量关上下更大的工夫。建议强化实地考察的力度，确保参展作品的质量。

在评比方式方面：

应该说，这次国家大展采用评委独立打分并在现场公布打分情况的评分规则，是目前国内最公开、最透明的办法，但如何真正做到令参展者满意，还需进一步完善和改进。建议采取“个人评”和“集体议”相结合的办法，对初选入围的大奖或金奖作品再合议一次，防止出现个别误评现象。

在展品的美化方面：

这次参展作品已注重配盆和盆面的精心处理，但国家大展和会员精品展中还有个别盆景没按要求配置几架，这种现象不应在今后的展览中发生。

在展场布置方面：展台、屏风布置已上了一个档次，但仍有美中不足，据参观者反映，屏风太小，灯光不均匀，影响了视觉和摄影效果。

首届中国盆景国家大展落户美丽富饶的中山古镇，虽然不是十全十美，但首届中国盆景国家大展终于闪亮登场，一个新的盆景城镇宣告诞生。它的成功举办，不仅拉动岭南区域盆景事业的飞速发展，而且给中国盆景界带来巨大而深远的影响。我们热切期待在广袤的中国大地上不断有新的盆景城镇的诞生，热切期待下一届中国盆景国家大展有更多高端精美的盆景面世，衷心祝愿中国盆景国家大展越办越好！

Show the Positive Energy of Pines' Plastic Arts

—Master Fan Shunli's Demonstration Performance of 2013 (Guzhen) China National Penjing Exhibition

展示松树造型艺术的正能量

——樊顺利大师在 2013（古镇）中国盆景国家大展上的现场制作表演

制作: 樊顺利
文: 胡光生
Processor: Fan Shunli
Author: Hu Guangsheng

制作者简介

樊顺利，中国盆景艺术大师、国际盆栽大师。现任中国盆景艺术家协会副会长、安徽省盆景艺术协会常务副会长、世界盆景石文化协会常务秘书长。

About the Creator

Fan Shunli, Chinese Penjing art master; International Penjing master. Now he is the vice president of China Penjing Artist Association; Executive vice-president of Anhui Penjing Artist Association; Executive secretary-general of World Bonsai Stone Culture Association.

图 1 黑松素材正面树相
The front side of the black pine material

这棵黑松素材经数年盆植，长势旺盛，从盆面至树梢达 1.65m，根基干径 30cm，主干略显弯曲，但角度不明显。由于没有过渡干，副枝干粗壮与第一出枝比例失调，使该树造型定势处理有一定难度。此外，其他枝位点也不是很理想，只能根据现有枝条的出枝点，通过调整主干的角度来做初步定势造型。

After several years, the black pine material is growing strong, whose height is up to 1.65m from the surface of pot to the top of the tree; the stem diameter at the base root of is 30cm; the trunk bent slightly, but the angle is not obvious. Because there is no excessive truck and the ratio between the second branch and main branch is imbalance, which made the tree modeling become more difficulty. In addition, the sites of other branches are not very ideal, so the processor should adjust the angle of the trunk to make preliminary modeling according to the existing sites of branches.

图 2 黑松素材背面树相
The back side of the black pine material

图 3、4 摘掉过密的松叶，疏理枝条，让树枝脉轮廓显现
Pick off the dense pine leaves, clear up the branches, and then make the contour come out

局部特写

图 5 依据枝条粗度选择不同型号的铝丝，对所有枝条进行绑扎
Choose different types of aluminums wire according to the thick of branches, after that tie up all the branches

局部特写

局部特写

局部特写

Show the Positive Energy of Pines' Plastic Arts

—Master Fan Shunli's Demonstration Performance of 2013 (Guzhen) China National Penjing Exhibition

图 6、7 对绑扎好的枝条进行布局定势，分别拿弯造型
Do layouts of the bind branches, and make them bend separately

图 8 定势后对枝托分条进行微调整
Readjust the branch after determined the position

局部特写

局部特写

图 9 中国盆景艺术家协会名誉会长鲍世骐与多位日本专家对樊顺利大师创作的松树进行点评;日本专家评说:我们观看制作过程很仔细,樊大师造型手法一流,这棵树做出了自然松树的韵味。这正展示出了松树盆景造型艺术的正能量
Mr. Bao Shiqi, honorary president of China Penjing Artist Association, and several Japanese experts make comment to the black pine. Japanese experts said we all look carefully and make sure that Master Fan has first-class modeling technique which made the pine naturally. It is show the positive energy in the pine Bonsai art modeling

图 10 日本专家向樊顺利大师提问:您是如何考量这棵黑松素材的?制作前是如何构想的?该树将来会怎么样?樊大师交出一张图给翻译女士转达在制作前的造型构思
Japanese experts asked Master Fan Shunli, How do you think of this black pine before the production? How is the future of this tree? Master Fan Shunli hand out a modeling design sketch to translator for explaining the conception before making

Show the Positive Energy of Pines' Plastic Arts

—Master Fan Shunli's Demonstration Performance of 2013 (Guzhen) China National Penjing Exhibition

图 11 制作前绘制的造型构思略图
The modeling design sketch before making

图 12 制作表演完成后的树姿，树势直立雄壮，充满阳刚之气。日后精心养护数年，反复调整造型，待枝干和谐后，定能成为松树盆景佳作
The gesture of the tree after performance erects imposing with full of masculinity. It must be a well pine Bonsai after stems being harmonious though several years carefully maintain and readjust the model

Making due to the Material

—Demonstration Performance of 2013 (Guzhen)China National Penjing Exhibition

随形就势，因材施艺

——2013（古镇）中国盆景国家大展现场制作表演

文：徐昊 Author: Xu Hao

能够成为2013(古镇)中国盆景国家大展现场制作表演嘉宾，我感到非常荣幸。按理说，制作表演应该自备或自选材料，才能更好地将自己的造型手法和创作理念展现给大家。但由于路途遥远，不方便自带材料，所以素材是既定的，只好随形就势、因材施艺地完成制作表演。

表演的素材是一棵体型较大的黑松桩坯，需要4个壮汉合力才能搬得动。这样大的素材，按照正常的制作速度，一个人大约需要一天多的时间才能完成从裁剪雕凿到整枝造型的全过程。要在规定的两个小时内完成制作任务，只能叫来数名学生，帮忙做摘叶及缠绕金属丝的工作。

图1 随着时间的推移，黑松的老叶已被摘去，露出可见的枝干，部分多余的粗枝也被去除
With the time passing, old leaves had been move, branches can be visible, and some extra thick branches had been removed

图 2、图 3 逐步剪除树冠中多余的枝条
Gradually cut off redundant branches in the canopy

I'm very honor to make a demonstration performance in 2013 (Guzhen) China National Penjing Exhibition. Normally, people should bring or choose materials by themselves, thus they can show their Modeling technique and creation concept better. Because the long journey for transfer materials, the demonstration performances use the existing material.

The material is a big black pine which need four strength man move. Make such a big material need more than one day from cutting, carving, to pruning branches. For completing the production task, several students help maker to pick leaves and twine wires

图 4 仔细审视主枝中每个小枝的取舍，剪除多余未用的小枝
Carefully decide the retention and remove of each sprig in the major branch, cut off the excess of unused sprig

图 5 选择顶部树枝的去留
Select the top branches

图 6 将留下的树枝按粗细逐级缠绕金属丝
Twine wires step by step for the rest branches

图7、图8向观众讲解制作的理念和制作要点
Explain the manufacture idea and key point to the audience

图9与观众互动，讨论该枝的取舍
Discuss how to cancel the branch with the audience

Making due to the Material

—Demonstration Performance of 2013 China (Guzhen) National Penjing Exhibition

图10主枝下方的这个树枝最终被剪除，留出空间，营造气韵之美。作者与观众讨论盆景造型的虚实关系以及树枝线条刚柔相间的应用手法
Cut off the branch below the main branch, and then create the beauty of artistic conception. Discuss the false or true relationship and the application of the soft and white branch lines in Penjing modeling with the audience

图11期间接受记者采访，回答关于盆景创作的一些问题
Accept an interview and answer some questions about Penjing creation

图 12 开始整理已扎上金属丝的树枝
Clear up the branches with wire fixed

图 13 将树枝调整到理想的位置
Adjust the branches to the desired location

图 14 整理顶部的枝条
Clear up the top of the branches

图 15 根据松树的自然特性，将顶枝做成自然横折、丰满结顶的模样
According to the natural properties of pine, make the top branch to cross breaks naturally and make the top plump

图 16 精心处理细枝的线条及位置
Clear up the thin branch's line and position carefully

图 17 用金属丝牵拉，进一步将粗枝调整到理想的位置，并将枝线作弯曲转折处理
Pull with wire, further adjust thick branch to the desired location, and twist it

图 18 逐渐整理好小枝
Clear up branchlets gradually

作品的根部强劲稳扎，主干粗壮，呈左斜之势，斜中寓曲，具有内在的力量感。作者根据素材的形质特点，将树枝作平展状布置，使之符合该树的生理特性。主枝强壮，曲直变化，向左凌空舒展，以此增强作品的气势和奇险的生境之美。主枝的下方以一背枝作点缀，增加了作品的立体透视感，同时展宕的树枝也平添了作品虚空的效果。

由于时间的关系，作品尚有许多未尽意之处，有待下一步继续完善，如桩节未曾处理；右边第一枝过度右展而影响到作品向左的气势，应当在下一步的制作中将其折回，或待近主干处的小枝稍粗壮后，将前段剪去等，逐步使之达到较为理想的效果。相信经过数年不断地完善和精心养护后，该作定能成为一件优秀的松树盆景作品。

图 19 在限定的时间内基本完成创作。回答观众的提问，讲析有待进一步改进之处
After finishing work in limited time, answer the audience's questions and analyze the further improved part

图 20 日本景道家世家二世须藤雨伯先生点评作品
Japanese Keido Iemoto(headmaster) II Mr Sudo Uhaku comment the work

图 21 作品完成图
The work after finishing

The root of this work is strong stability; the main stem is thick and solid, oblique have curve, with inner strength. According to the material's characteristics, the producer made flat shape layout to match the physiological characteristics of the tree. The main branch is strong and with grasp of change and volley stretch to the left to strengthen the work's vigor and the beauty of the habitat. At the bottom of the main branch decorated with a dorsal branch, which increased the sense of three-dimensional perspective and stretched branches show a void effect.

Because of the short time, there are many disadvantages which should continue to improve, such as unhandled pile section; the excessive right branch effect the work's left imposing manner, thus for reaching the ideal effect the next step should turn it back or cut down the top section when the sprig near main branch turn thicker. After several years' continuous growing and careful maintenance, it must be an excellent pine Penjing work.

"岁月如歌" 真柏
高 98cm 宽 130cm
刘桂球藏品

The Annual Award Works China Penjing Member Exhibition Appreciation of 2013

2013中国盆景会员展年度大奖作品欣赏

三角梅 高 120cm 宽 100cm
劳寿权藏品

刺柏 高 120cm 宽 170cm 马建中藏品

“清涛雅韵” 五针松 高 95cm
宽 100cm 沈水泉藏品

金奖作品欣赏
2013中国盆景会员展
2013

Gold Award Works Appreciation of 2013 China Penjing Member Exhibition

“涅槃” 真柏 高 88cm
宽 160cm 陈光华藏品

“罗汉叠翠” 短叶罗汉松
高 120cm 宽 180cm 李正银藏品

山橘 高 128cm 李仕灵藏品

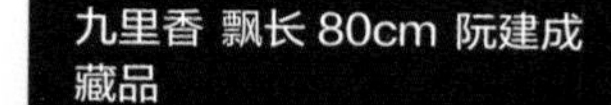
九里香 飘长 80cm 阮建成
藏品

“谦谦君子” 山松 高 100cm 彭盛材藏品

九里香 高 120cm 宽 90cm
肖庚武藏品

“沐雨栉风” 真柏 高 100cm
宽 110cm 朱昌圣藏品

Silver Award Works Appreciatio

China Pen

Member Exhibition

“山乡吉临” 山橘 高 82cm 宽 100cm 程煜涵藏品

三角梅 飘长 100cm 何伟源藏品

榕树 高 130cm 宽 200cm 柯成昆藏品

空叠彩” 真柏 高 120cm
20cm 黎炽雄藏品

“东方神韵” 刺柏
高 93cm 王炘藏品

“峭壁攒峰千万枝”雀梅（悬崖）高 120cm 吴成发藏品

山橘 高 95cm 黄继涛藏品

黄杨 高 100cm 宽 100cm
陈国健藏品

2013 中国盆景会员展银奖作品欣赏

Silver Award Works Appreciation of
China Penjing
Member Exhibition

公 飘长 80cm 李金成藏品

真柏 高 75cm 宽 110cm
陈明兴藏品

“雅赞” 罗汉松 高 95cm
宽 130cm 关山藏品

玉叶 高 90cm 宽 130cm 张金回藏品

2013 中国盆景会员展银奖作品欣赏

“咏将”罗汉松 高 108cm 宽 170cm 关山藏品

九里香 高 120cm 林伟栈藏品

枫树 高 118cm 宽 110cm 袁心义藏品

相思 高 120cm 宽 110cm 陈光明藏品

刺柏 高 85cm 宽 110cm 马建中藏品

真柏 高 120cm 宽 130cm 刘永洪藏品

“榕韵” 榕树 高 120cm 洪容兴藏品

“礼”真柏 高 115cm 宽 75cm 吴国庆藏品

2013

Bronze Award Works Appreciation of 2013 China Penjing Member Exhibition

中国盆景

会员展铜奖作品欣赏

簕杜鹃 高 90cm 梁志坚藏品

“翠影” 真柏 高 110cm 吴成发藏品

“福荫华夏” 福建茶 高 104cm 劳杰林藏品

"天宫胜景" 清香木附石 高 130cm 陈迎凯、何兆良藏品

黑松 飘长 50cm
欧阳国耀藏品

"水是故乡美"过桥榕
高 76cm 吴松恩藏品

"柏芽扶绿" 真柏
高 87cm 宽 130cm 李晓波藏品

"灵苗绝伦" 红牛 高 115cm
吴成发藏品

"如意" 九里香 高 66cm 宽 97cm
徐伟华藏品

山松 高 70cm 麦永强藏品

"廊桥遗梦" 雀梅
高 70cm 宽 100cm
刘学武藏品

"仙女飘然下凡间" 山松
高 92cm 林连生藏品

"南珠风情" 六角勁
高 90cm 陈家勁藏品

"南国乡情" 榕树 125cm
李建藏品

"松韵" 五针松 高 115cm
宽 140cm 管银海藏品

"魔界" 万年阴
飘长 130cm 吴成发藏品

相思 飘长 80cm
欧阳铭初藏品

Bronze Award Works Appreciation of 2013 China Penjing Member Exhibition

"绿之恋" 榆树 高 80cm 宽 131cm 李晓波藏品

"夫唱妇随" 刺柏 高 75cm 宽 50cm 孙友祥藏品

"游龙戏凤" 六角榕 高 150cm 宽 150cm 暨佳藏品

山松 高 100cm 何焯光藏品

"早春" 榕树 高 96cm 宽 115cm 柯成昆藏品

中国盆景 会员展铜奖作品欣赏

"寒宫舒袖" 真柏 高 110cm 宽 140cm 魏积泉藏品

"泰山精魂" 榕树 高 127cm 宽 165cm 徐伟华藏品

"小娇" 榕树 高 80cm 宽 110cm 陈文辉藏品

台湾真柏 高 75cm 宽 170cm 吴明选藏品

"情怀江绪" 水旱相思 高 70cm 何长洪藏品

博兰 高 150cm 宽 80cm 卢炳权藏品

“荣辱与共”真柏 高 50cm 长 65cm 陈冠军藏品

“三口之乐沐春风” 榆树
高 110cm 长 128cm 陈再米藏品

“松籁” 黑松 高 80cm
宽 80cm 史佩元藏品

榕树 高 116cm 宽 170cm
庄振良藏品

黑松 高 90cm 宽 118cm
陈文君藏品

六月雪 高 65cm
宽 80cm
袁心义藏品

榕树 高 120cm 宽 130cm 陈光明藏品

“飞云横渡” 黄杨 高 50cm
宽 135cm 王礼宾藏品

“双松迎客” 五针松
高 120cm 宽 120cm
孙筱良藏品

“大地情深” 石榴 高 120cm
宽 150cm 王鲁晓藏品

“盛” 罗汉松 高 115cm
宽 145cm 刘桂球藏品

“香果丽拉” 美国卡丽沙
高 93cm 宽 138cm 林国富藏品

2013 中国盆景艺术家协会会员盆景精品展评分表

Assessment Scoring Form of China Penjing Member Exhibition of China Penjing Artists Association

评比编号	参展者	地区	题名	树种	拉丁名	规格(cm)	奖项	得分	名次	谢克英	王选民	徐昊	曾文安	陆志伟	罗传忠	徐闻	总分	去掉最高分	去掉最低分	有效总分
154	刘桂球	湖北	岁月如歌	真柏	*Juniperus chinensis* var. *sargentii*	高 98 宽 130	大奖	91.60	1	85	88	93	92	93	96	92	639	96	85	458
077	劳寿权	广东		三角梅	*Bougainvillea spectabilis*	高 120 宽 100	金奖	91.20	2	96	88	77	86	93	95	94	629	96	77	456
158	李正银	广西	罗汉叠翠	短叶罗汉松	*Podocarpus macrophyllus*	高 120 宽 180	金奖	90.60	3	95	88	78	86	91	95	93	626	95	78	453
186	马建中	浙江		刺柏	*Juniperus formosana*	高 120 宽 170	金奖	87.80	4	93	86	93	68	89	78	93	600	93	68	439
123	彭盛材	广东	谦谦君子	山松	*Pinus massoniana*	高 100	金奖	86.80	5	92	68	87	88	93	92	75	595	93	68	434
147	陈光华	湖北	涅槃	真柏	*Juniperus chinensis* var. *sargentii*	高 88 宽 160	金奖	86.80	5	87	88	87	86	93	86	70	597	93	70	434
003	阮建成	广东		九里香	*Murraya exotica*	飘长 80	金奖	85.00	7	85	68	91	88	91	87	74	584	91	68	425
187	沈水泉	浙江	清涛雅韵	五针松	*Pinus parviflora*	高 95 宽 100	金奖	85.00	7	92	88	92	78	93	67	75	585	93	67	425
111	肖庚武	广东		九里香	*Murraya exotica*	高 120 宽 90	金奖	84.40	9	69	88	83	88	81	85	85	579	88	69	422
057	李仕灵	广东		山橘	*Fortunella hindsii*	高 128	金奖	83.80	10	93	78	78	92	93	78	78	590	93	78	419
061	朱昌圣	大观园	沐雨栉风	真柏	*Juniperus chinensis* var. *sargentii*	高 100 宽 110	金奖	83.60	11	84	75	92	92	86	81	70	580	92	70	418
071	关山	大观园	咏将	罗汉松	*Podocarpus macrophyllus*	高 108 宽 170	银奖	83.00	12	86	88	66	88	73	82	86	569	88	66	415
019	吴成发	香港	峭壁攒峰千万枝	雀梅（悬崖式）	*Sageretia theezans*	高 120	银奖	82.80	13	75	88	78	66	91	82	91	571	91	66	414
078	黄继涛	广东		山橘	*Fortunella hindsii*	高 95	银奖	82.60	14	76	96	95	88	78	76	70	579	96	70	413
108	关山	大观园	雅赞	罗汉松	*Podocarpus macrophyllus*	高 95 宽 130	银奖	82.60	14	93	93	67	92	78	68	82	573	93	67	413
021	洪容兴	广东	榕韵	榕树	*Ficus microcarpa*	高 120	银奖	82.40	16	67	93	68	76	86	91	91	572	93	67	412
117	林伟栈	广东		九里香	*Murraya exotica*	高 120	银奖	81.60	17	85	68	77	92	75	86	85	568	92	68	408
035	刘永洪	四川		真柏	*Juniperus chinensis* var. *sargentii*	高 120 宽 130	银奖	81.20	18	75	76	94	76	89	87	78	575	94	75	406
130	张金回	福建	绿树明珠	玉叶		高 90 宽 130	银奖	81.20	18	93	76	77	95	78	76	82	577	95	76	406
016	李金成	广东		山松	*Fortunella hindsii*	飘长 80	银奖	80.40	20	78	78	67	88	83	96	75	565	96	67	402
127	吴国庆	福建	礼	真柏	*Juniperus chinensis* var. *sargentii*	高 115 宽 75	银奖	80.20	21	75	95	68	95	76	85	70	564	95	68	401

（续）

评比编号	参展者	地区	题名	树种	拉丁名	规格(cm)	奖项	得分	名次	谢克英	王选民	徐昊	曾文安	陆志伟	罗传忠	徐闻	总分	去掉最高分	去掉最低分	有效总分
184	陈明兴	江苏		真柏	*Juniperus chinensis* var. *sargentii*	高 75 宽 110	银奖	78.80	22	76	96	92	78	70	67	78	557	96	67	394
014	黎炽雄	广东	凌空叠彩	真柏	*Juniperus chinensis* var. *sargentii*	高 120 宽 120	银奖	78.60	23	76	68	68	68	93	95	88	556	95	68	393
169	马建中	浙江		刺柏	*Juniperus formosana*	高 85 宽 110	银奖	78.20	24	75	78	91	72	75	67	92	550	92	67	391
121	陈光明	广东		相思	*Acacia confusa*	高 120 宽 110	银奖	78.00	25	78	68	78	78	78	81	78	539	81	68	390
002	程煜涵	广东	山乡吉临	山橘	*Fortunella hindsii*	高 82 宽 100	银奖	77.60	26	67	76	78	78	78	78	78	533	78	67	388
133	柯成昆	福建		榕树	*Ficus microcarpa*	高 130 宽 200	银奖	77.60	26	85	76	67	68	75	97	84	552	97	67	388
026	王炘	浙江	东方神韵	刺柏	*Juniperus formosana*	高 93 长 102 宽 85	银奖	77.40	28	68	78	88	92	75	78	68	547	92	68	387
056	何伟源	广东		三角梅	*Bougainvillea spectabilis*	飘长 100	银奖	77.40	28	78	76	68	82	73	78	86	541	86	68	387
163	陈国健	福建		黄杨	*Buxaceae*	高 100 宽 100	银奖	77.40	28	78	78	85	88	68	78	68	543	88	68	387
038	袁心义	浙江		枫树	*Aceraceae*	高 118 高 110	银奖	77.20	31	76	78	77	92	83	67	72	545	92	67	386
103	梁志坚	广东		簕杜鹃	*Bougainvillea spectabilis*	高 90	铜奖	77.20	31	66	78	85	78	75	86	70	538	86	66	386
018	李晓波	浙江	柏芽扶绿	真柏	*Juniperus chinensis* var. *sargentii*	高 87 宽 130	铜奖	76.80	33	74	68	88	75	89	78	69	541	89	68	384
022	管银海	浙江	松韵	五针松	*Pinus parviflora*	高 115 宽 140	铜奖	76.80	33	77	85	78	68	88	68	76	540	88	68	384
072	林连生	广东	仙女飘然下凡间	山松	*Pinus massoniana*	高 92	铜奖	76.80	33	85	78	69	68	70	82	87	539	87	68	384
100	麦永强	广东		山松	*Pinus massoniana*	高 70	铜奖	76.80	33	66	76	78	68	78	84	91	541	91	66	384
191	陈文辉	福建	小娇	榕树	*Ficus microcarpa*	高 80 宽 110	铜奖	76.80	33	78	68	88	92	75	67	75	543	92	67	384
084	徐伟华	广西	如意	九里香	*Murraya exotica*	高 66 宽 97	铜奖	76.60	38	76	86	78	66	68	91	75	540	91	66	383
031	陈再米	浙江	三口之乐沐春风	榆树	*Ulmus pumila*	高 110 长 128	铜奖	76.40	39	69	68	91	78	85	78	72	541	91	68	382
042	孙筱良	浙江	双松迎客	五针松	*Pinus parviflora*	高 120 宽 120	铜奖	76.40	39	75	88	63	78	78	76	75	533	88	63	382
045	吴成发	香港	翠影	真柏	*Juniperus chinensis* var. *sargentii*	高 110	铜奖	76.40	39	68	88	67	68	88	94	70	543	94	67	382
076	何焯光	广东		山松	*Pinus massoniana*	高 100	铜奖	76.20	42	85	68	69	66	73	86	91	538	91	66	381
008	吴松恩	广东	水是故乡美	榕树	*Ficus microcarpa*	高 76	铜奖	76.00	43	68	76	92	86	68	68	82	540	92	68	380
134	吴明选	福建		台湾真柏		高 75 宽 170	铜奖	75.80	44	68	96	78	88	75	68	70	543	96	68	379
151	林国富	福建	香果丽拉	美国卡丽沙		高 93 宽 138	铜奖	75.80	44	85	68	76	86	73	75	70	533	86	68	379
079	李建	广东	南国乡情	榕树	*Ficus microcarpa*	高 125	铜奖	75.60	46	76	68	66	75	88	76	83	532	88	66	378
101	徐伟华	广西	泰山精魂	榕树	*Ficus microcarpa*	高 127 宽 165	铜奖	75.60	46	85	68	82	68	75	68	87	533	87	68	378
180	袁心义	浙江		六月雪	*Serissa japonica*	高 65 高 80	铜奖	75.60	46	85	78	85	72	73	68	70	531	85	68	378

（续）

评比编号	参展者	地区	题名	树种	拉丁名	规格(cm)	奖项	得分	名次	谢克英	王选民	徐昊	曾文安	陆志伟	罗传忠	徐闻	总分	去掉最高分	去掉最低分	有效总分
053	陈家劲	广东	南珠风情	六角劲		高 90	铜奖	75.40	49	78	68	67	68	85	91	78	535	91	67	377
135	柯成昆	福建	早春	榕树	*Ficus microcarpa*	高 96 宽 115	铜奖	75.40	49	78	83	86	66	70	68	78	529	86	66	377
082	刘学武	广西	廊桥遗梦	雀梅	*Sageretia theezans*	高 70 宽 100	铜奖	75.20	51	76	78	87	66	75	73	74	529	87	66	376
098	何长洪	广东	情怀江绪	相思	*Celtis sinensis*	高 70	铜奖	75.20	51	69	88	68	68	85	82	72	532	88	68	376
005	欧阳铭初	广东		相思	*Celtis sinensis*	飘长 80	铜奖	75.00	53	76	78	68	68	85	88	68	531	88	68	375
066	吴成发	香港	灵苗绝伦	红牛	*Scolopia chinensis*	高 115	铜奖	74.60	54	84	68	68	62	76	83	78	519	84	62	373
137	庄振良	福建		榕树	*Ficus microcarpa*	高 116 宽 170	铜奖	74.60	54	85	68	81	68	70	78	76	526	85	68	373
138	魏积泉	福建	寒宫舒袖	真柏	*Juniperus chinensis* var. *sargentii*	高 110 宽 140	铜奖	74.60	54	76	78	76	72	71	68	82	523	82	68	373
167	王鲁晓	山东	大地情深	石榴	*Punica granatum*	高 120 宽 150	铜奖	74.60	54	86	86	75	72	70	68	70	527	86	68	373
175	孙友祥	浙江	夫唱妇随	刺柏	*Juniperus formosana*	高 75 宽 50	铜奖	74.60	54	66	76	86	86	73	68	70	525	86	66	373
044	陈文君	浙江		黑松	*Pinus thunbergii*	高 90 宽 118	铜奖	74.40	59	78	78	77	72	75	68	70	518	78	68	372
064	劳杰林	广东	福萌华夏	福建茶	*Carmona microphylla*	高 104 宽 76	铜奖	74.40	59	92	68	67	76	73	77	78	531	92	67	372
148	史佩元	江苏	松籁	黑松	*Pinus thunbergii*	高 80 宽 80	铜奖	74.20	61	76	78	67	85	70	78	69	523	85	67	371
183	陈冠军	江苏	荣辱与共	真柏	*Juniperus chinensis* var. *sargentii*	高 50 长 65	铜奖	74.20	61	66	78	75	75	75	68	78	515	78	66	371
063	陈迎凯 何兆良	广东	天宫胜景	清香木附石		高 130	铜奖	74.00	63	78	75	66	68	71	78	88	524	88	66	370
132	陈光明	广东		榕树	*Ficus microcarpa*	高 120 宽 130	铜奖	73.80	64	76	68	78	72	83	73	70	520	83	68	369
139	王礼宾	福建	飞云横渡	黄杨	*Buxus sinca*	高 50 宽 135	铜奖	73.80	64	78	68	78	88	75	68	70	525	88	68	369
010	欧阳国耀			黑松	*Pinus thunbergii*	飘长 50	铜奖	73.60	66	68	68	68	82	78	76	78	518	82	68	368
083	吴成发	香港	魔界	万年阴		飘长 130	铜奖	73.60	66	66	68	69	78	91	68	85	525	91	66	368
112	卢炳权	广东		博兰	*Ponamella fragilia*	高 80 宽 150	铜奖	73.60	66	83	68	69	75	73	68	91	527	91	68	368
155	刘桂球	湖北	盛	罗汉松	*Podocarpus macrophyllus*	高 115 宽 145	铜奖	73.40	69	78	95	65	78	73	68	70	527	95	65	367
043	李晓波	浙江	绿之恋	榆树	*Ulmus pumila*	高 80 宽 131	铜奖	73.20	70	69	92	78	68	73	68	78	526	92	68	366
060	暨佳	广东	游龙戏凤	六角榕	*Ficus microcarpa*	高 150 宽 150	铜奖	73.00	71	78	68	65	68	83	68	85	515	85	65	365

清早期 炉钧釉圆飘口圈足盆 直径38.5cm 高20.5cm 杨贵生藏品
Early Qing Dynasty Robin's Egg with Round Overhanging-Edge Pot. Diameter: 38.5cm, Height: 20.5cm.
Collector: Yang Guisheng

清早期 乌泥圆八卦纹兽脸足盆 直径39.5cm 宽11.3 cm 申洪良藏品
Early Qing Dynasty Dark Clay with Trigram Pattern and Feast Face Leg Pot. Diameter: 39.5cm, Width: 11.3cm.
Collector: Shen Hongliang

清早期 乌泥树桩型盆 长38cm 宽32cm 高15.5cm 杨贵生藏品
Early Qing Dynasty Dark Clay in Tree Stump Shape Pot. Length: 38cm, Width: 32cm, Height: 15.5cm. Collector: Yang Guisheng

清初 乌泥圆鼓钉盆 直径37cm 高8 cm 申洪良藏品
Early Qing Dynasty Dark Clay with Round Nail Shape Pot. Diameter: 37cm, Height: 8cm. Collector: Shen Hongliang

清初 紫泥长方飘口底线墨彩盆 长54cm 宽33cm 高29.5cm 杨贵生藏品
Early Qing Dynasty Purple Clay with Chinese Painting and Overhanging-Edge Rectangular Pot.
Length: 54cm, Width: 33cm, Height: 29.5cm. Collector: Yang Guisheng

清初 乌泥海棠飘口上下带线盆 长48cm 宽39.5cm 高26cm申洪良藏品
Early Qing Dynasty Dark Clay with Two Stripe and Crabapple Shape Overhanging-Edge Pot.
Length: 48cm, Width: 39.5cm, Height: 26cm. Collector: Shen Hongliang

明末 乌泥长方直壁上下带线盆 长45.5cm 宽29cm 高17.5cm 杨贵生藏品
Late-Ming Dynasty Dark Clay with Two Lines Rectangular Pot. Length: 45.5cm, Width: 29cm, Height: 17.5cm.
Collector: Yang Guisheng

明末 乌泥正方飘口段足盆 长 38cm 宽 38cm 高 25cm 申洪良藏品

清早期 乌泥树桩型盆 长 38cm 宽 32cm 高 15.5cm 杨贵生藏品

清早期 乌泥圆八卦纹兽脸足盆 惠逸公 直径 39.5cm 宽 11.3 cm 申洪良藏品

清初 紫泥长方飘口底线墨彩盆 长54cm 宽33cm 高29.5cm 杨贵生藏品

（左）清早期 蓝釉长方飘口开光花足盆 长 31cm 宽 20.5cm 高 16cm 申洪良藏品
（右）清初 乌泥长方抚角直壁双带线盆 长 33.7cm 宽 23cm 高 9cm 申洪良藏品

清代 红泥椭圆袋式盆 徐友泉制 长 46cm 宽 34cm 高 14.5cm 关山藏品

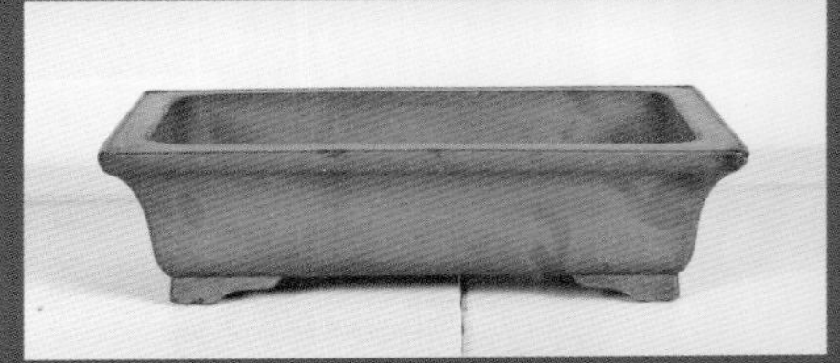

清初 红泥长方飘口段足盆 长 36cm 宽 23.5cm 高 10cm 申洪良藏品

2013 中国盆景会员展之古盆欣赏

Ancient Pot Appreciation of 2013 China Penjing Member Exhibition

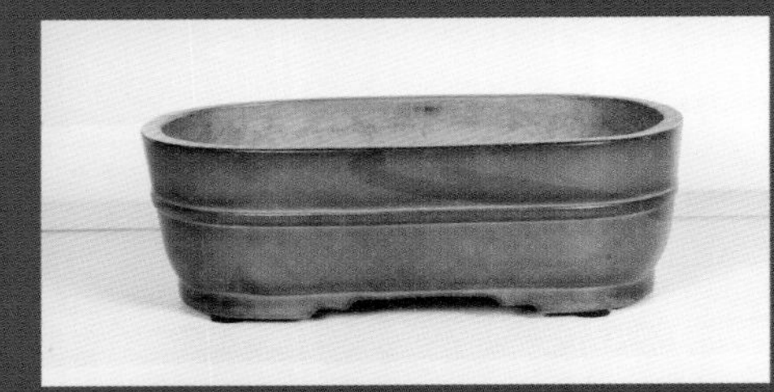

清中期 白泥椭圆中带线盆 长 38cm 宽 25cm 高 12cm 申洪良藏品

明末 红泥长方飘口玉绿底线切足盆 长 68cm 宽 39cm 高 23.8cm 申洪良藏品

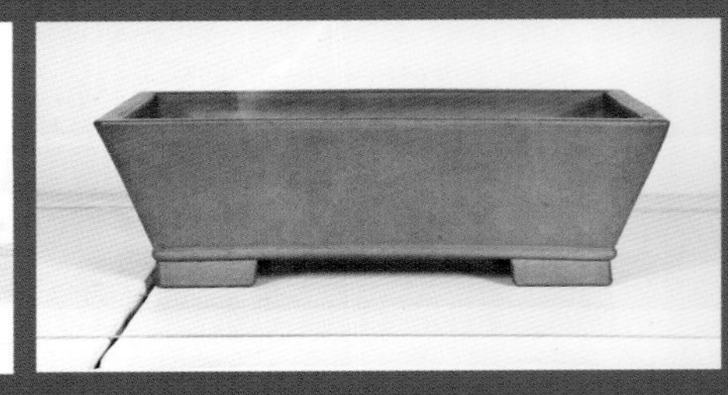

清早期 乌泥直壁底线切足盆 长 44cm 宽 26.5cm 高 14cm 杨贵生藏品

清早期 乌泥正方直壁底线墨彩盆 长 33.5cm 高 26cm 杨贵生藏品

清早期 乌泥大瓜轮底线圈足盆 直径 63cm 高 35cm 杨贵生藏品藏品

清早期 乌泥椭圆直壁盆 长 36cm 宽 27.5cm 高 13cm 杨贵生藏品

清中期 乌泥外缘切足四方盆 长 38cm 宽 38cm 高 27 cm 金文明藏品

明末 乌泥椭圆飘口双线盆 长 62cm 45.5cm 22.5cm 杨贵生藏品

清初 乌泥圆鼓钉盆 直径 37cm 高 8 cm 申洪良藏品

清早期 炉钧釉圆飘口圈足盆 直径 38.5cm 高 20.5cm 杨贵生藏品

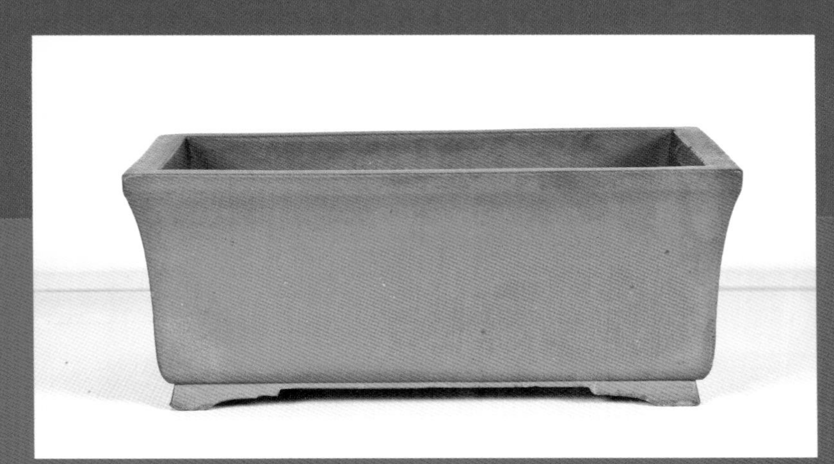

清早期 红泥飘口段足盆“为善最乐”长 36cm 宽 22cm 高 15cm 申洪良藏品

清早期 红泥圆飘口上下带线圈足盆 直径 52cm 高 29 cm 杨贵生藏品

清中期 紫泥正方飘口连足盆 长 38cm 高 22.5cm 申洪良藏品

清初 乌泥海棠飘口上下带线盆 长 48cm 宽 39.5cm 高 26cm 申洪良藏品

（左）清初 红泥长方飘口兽面足盆 长 35cm 宽 23cm 高 13cm 申洪良 藏品
（右）清中期 朱泥六角飘口底线段足盆 长 28cm 宽 25cm 高 16cm 申洪良藏品

清代 朱泥椭圆袋式盆 陈文卿制 长 42cm 宽 29cm 高 9cm 申洪良藏品

民国 蓝釉长方袋式盆 陈恒丰制 长26cm 宽17.5cm 高 7.5 cm 杨贵生藏品

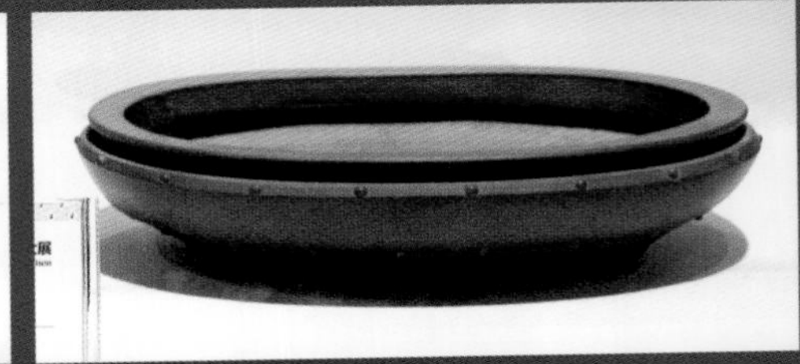

民国 紫泥椭圆袋式鼓钉盆 长 38.5cm 宽 27.5cm 高 8cm 杨贵生藏品

清中期 紫泥外缘云足长方盆 长 50cm 宽 30cm 高 10.5cm 金文明藏品

2013 中国盆景会员展之古盆欣赏

Ancient Pot Appreciation of 2013 China Penjing Member Exhibition

民国 初粉红泥圆刻绘盆一对
直径 41cm 高 22cm 金文明藏品

民国 粉段泥圆刻绘盆一对
直径 41cm 高 22.5cm 金文明藏品

民末 黄釉长方贴字盆 长 40.7cm
宽 27cm 高 27cm 杨贵生藏品

民国 紫泥八角袋式僧帽盆
直径 33cm 高 14cm 杨贵生藏品

现代 朱泥正方飘口抽角段足盆
长 25cm 宽 25cm 高 18.5cm
申洪良藏品

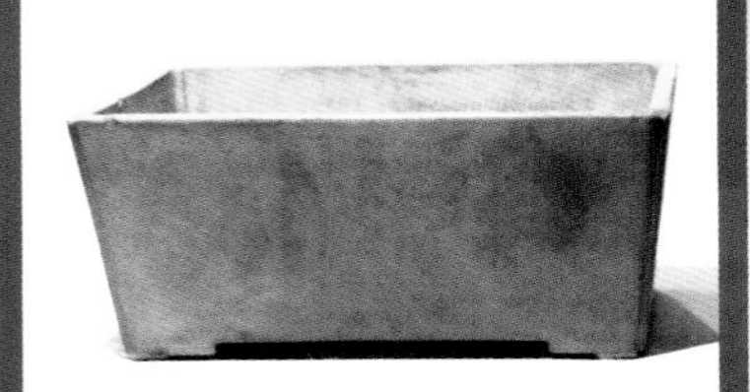

清中期 白泥长方直壁抽角切足盆
长 35cm 宽 24.5cm 高 16cm
杨贵生藏品

清 朱泥小瓜轮渣斗盆 陈文居制
直径 29.5cm 高 20 cm 申洪良藏品

清早期 紫泥梨皮圆飘口口线圈足盆
直径 52cm 高 25.5cm 杨贵生藏品

清中期　紫泥黄泥绘外缘线奎连
足六方盆长 46cm 宽 46cm
高 25cm 金文明藏品

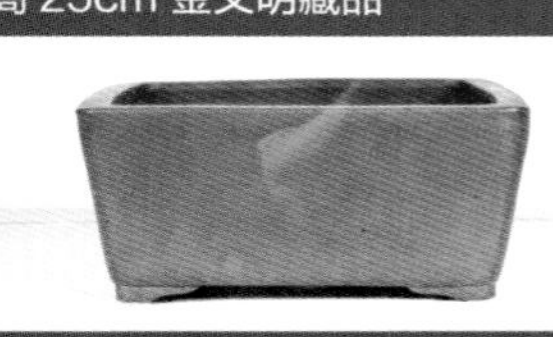

清初 红泥正方直壁段足盆
长 31cm 宽 13cm 申洪良藏品

清早期 大红袍桃花泥墨彩山水画盆 长69cm 宽40cm 高25cm 金文明藏品

清初 乌泥长方飘口口线段足盆
长 45cm 宽 29cm 高 18cm
杨贵生藏品

清中期 乌泥玉沿下纽切足长方盆
长 46.5cm 宽 31cm 高 18cm
金文明藏品

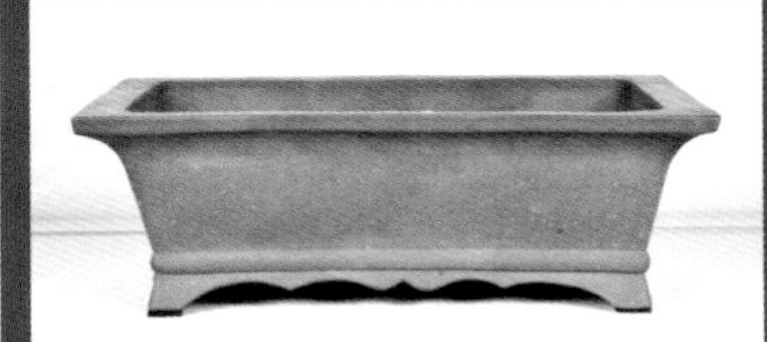

清初 紫泥梨皮长方飘口底线花足盆 长
48.5cm 宽 31.5cm 高 14cm
杨贵生藏品

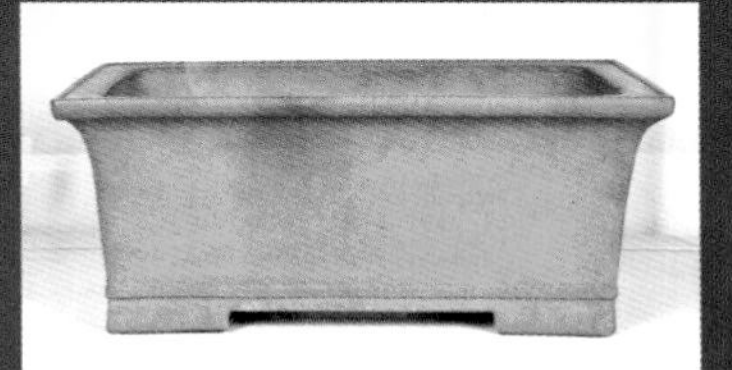

清早期 桃花外缘下玉带连足长方盆
长 61cm 宽 32cm 高 21.5cm
金文明藏品

明末 乌泥长方直壁上下带线盆 长 45.5cm
宽 29cm 高 17.5cm 杨贵生藏品

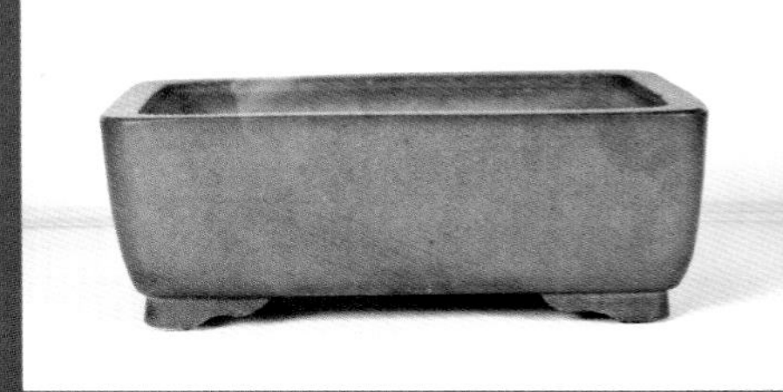

清早期 乌泥长方直壁段足盆 长 35cm 宽 25cm
高 13cm 杨贵生藏品

民国 粉段泥长方飘口刻绘盆 长 56.8cm
宽 34.5cm 高 18cm 申洪良藏品

个人简介

刘洪，中国盆景艺术家协会名誉副会长。

可圈可点——我看中国盆景国家大展

Commendable–I see China National Penjing Exhibition

文：刘洪 Author：Liu Hong

这次中国盆景国家大展有5点原因吸引我前来参加，一是由地方政府参与主办；二是国家大展的规模气势；三是独特的办展思路；四是一流的运作形式；五是有力度的宣传。

相比以前参加过的展览，总括而论，此次中国盆景国家大展具有汉唐"恢弘壮阔"的气魄和令人耳目一新的感受，体现在办展的规模、布展的设计、媒体的宣传力度、众多国内外有影响力的嘉宾参与、参展作品风格的多样性与高水准五个层面上。例如，此次大展的办展思路非常开阔，融汇国际、国内大展之长，邀请了众多国际、国内有影响力的盆景艺术工作者，为国际、国内的盆景艺术家提供了展现、交流作品的宽阔平台；在布展的设计上，将中国的传统文化与现代的具象艺术有机地融入布展思路，彰显了中华文化的丰厚底蕴。如大展的LOGO设计可谓独具匠心。取汉字"鼎"的意、形，将二者巧妙结合。"鼎"作为中国古代传统的礼器，有尊崇之意，表达了主办方对来自海内外的盆景艺术家的礼遇与热忱之意；鼎上之"目"的设计巧妙地昭示了大展的地域特征和我国盆景制作的悠久历史以及对未来盆景事业发展生机盎然的美好祝愿。而"鼎"之形又恰似一幅盆景作品。此外，将传统文化与现代的传媒有机结合，向世人展示现代中国的独特风采和盆景作品风格的多样性也是这次大展的一大特点。与相关国外展览不同之处在于政府的介入，而国外的盆景大展往往由民间机构举办，且宣传力度远不及国内展览。

此次展览可圈可点之处颇多，例如，参展的盆景作品大都具有相当的水准。表现在：参展作品的造型设计特色纷呈。即具有传统中国盆景流派的作品展现，又可以看到近些年国内不同区域作品的推陈出新；还可以看到一些中西合璧作品的探索。作品所涉及的树种也比较丰富。美中不足的是，这次大展中小型盆景的数量相对不足，使人们失去了欣赏“小中见大”之美的意趣，同时，山水盆景的数量严重不足。山水盆景可以说是中国盆景独有的形式，它最能展现中国文化的精髓，且愈加受到世人的喜爱，建议今后设专门的栏目探讨。此届古盆、赏石展品品质较高，唯品种单一，两者都面临进一步挖掘问题。举例而论，中国历代古盆官窑产品均有存世，每年春秋两次大拍均有上述产品推出，足可见民间收藏市场可观；同理，中国地域辽阔，赏石资源丰富，品类众多。特别是近年来，鉴石、藏石、赏石活动日趋火爆，建议协会应引起足够重视。

此届盆景大展虽然特色纷呈，但能够称得上创新的盆景作品我还没有发现。我认为盆景创新应包含两点：一是盆景理念的创新，二是盆景技法的创新。

依据现有评比规则，评比结果较令人满意。评比规则的设立应包含两方面内容：一是权威评比机制，二是民意评比机制，两者相互依托，不可或缺。9 位评委和监委组成的评委团队的设立应属于权威评比机制，但我认为还不够，还应建立一套民意评比机制，只有这样才能比较全面地体现公平、公正的原则。为此，我赞成评判盆景采用评分表的形式。

中国盆景年度之夜的晚会气氛雍容、热烈、祥和，亮点颇多。举办晚会既是表达东道主待客之道，也是展示主办方民族文化的良好时机，同时，还是对外文化交流的重要窗口。古人云“有朋自远方来，不亦悦乎”正是此意。比如沙画表演、模特表演、贵宾热舞等都很有亲和性，方便了来宾之间的交流。

本届大展筹备工作组织的颇具专业性。无论会场外接待、食宿、交通的层面，还是会场内各项交流活动的层面皆组织得近乎完美，无可挑剔。特别需要指出的是，工作人员服务热情、周到，给人一种宾至如归之感。唯一尚待改进的地方是应协调好短暂与永恒的关系。艺术的精髓是倡导永恒，倡导时间的久远。大展是短暂的，但它所秉承的精神是永恒的。特别是如此重要的大展，面对海内外的众多嘉宾，如撰写一个“古镇大展宣言”，言简意赅地阐述大展的精神，将会使来宾印象更为深刻；也可以铭刻在一块景观石上，永久纪念，令人回味。

作为一个盆景人，这是一个期待已久的大展。它以全方位的角度向世界展示了古老的文明古国弘扬其灿烂的民族文化的雄心、信心；展示了当代中国雄厚的经济实力；也展示了中国盆景人海纳百川、兼容并蓄的胸怀。这是一个多元化的时代，机遇与挑战共存，困惑与喜悦同在。中国盆景就如同 20 世纪初中国绘画一样会遇到各种文化形态的交融、碰撞，也会出现流派纷呈、百家争鸣的局面。言谈至此，我忽然记起 20 世纪初胡适先生告诫青年的名言：“大胆假设，小心求证”，此言至今仍有积极的意义，它告诫我们在盆景文化领域，面对各种外来文化思潮，既要兼收并蓄，又要冷静思考。

个人简介

曾尔恩，中国盆景艺术家协会名誉常务副会长。

浅谈 中国鼎 2013（古镇）中国盆景国家大展

文：曾尔恩 Author: Zeng Er'en

2013（古镇）中国盆景国家大展吸引我的地方在于入围难。有别于一般的展览，这次的国家大展和会员精品展，还有古盆和赏石展，都是经过精挑细选的，能获得参展资格的，必然是优秀作品，具有一定的代表性。能够在展览上一次欣赏到这么多好的作品，实在是一次难得的学习机会。

作为这个领域的新人，我所参加过的展览并不多，这次展览展品的水平毋庸置疑。日本人在做人做事方面认真细致，他们专注于松柏类盆景，对盆、摆件、几架的配衬极其注重，这种对艺术的态度非常值得我们学习。中国是一个多民族的国家，在美食、绘画、武术、舞蹈、戏剧、语言等方面都有各自的特色，这是其他国家及地区无法与之相比的。在盆景方面，树的品种非常丰富，表达的手法也各有不同，题材更是取之不尽，我们可以将中国几千年的历史文化表现在盆景上。所以国与国之间，地区与地区之间，不应用同一定义来衡量。

这次国家大展的名称非常有意义，"中国鼎"，正如中国盆景艺术家协会会长苏放先生所说："鼎"这个中国古代最重要的传国礼器能最恰当地代表一个国家级的最高奖项应有的王者般的荣耀，也预示着中国的盆景进入鼎盛时期。而这次的中国盆景年度先生已经诞生，作为女性，希望在不久的将来，中国盆景女士也会相继出现。从而吸引更多的女性和年轻人投入到这个领域，让世界的盆景艺术承传下去。

这次的参展的作品，除了我们以往在其他展览上、《中国盆景赏石》上所见到的一些佳作之外，还出现了很多新作，很多展品也是我们平常难得一见的，参观者无不感到惊讶。无论大型的、小型的，都有合适的几架配衬，树的造型与枝爪也更加成熟。岭南盆景有许多都带叶出现，让人们一方面欣赏到树的造型，细细的树叶或果实也让大家感受到树木的生机勃勃。再有就是奇、怪的展品有所增加，这些都是作者运用大自然所提供的素材，以最自然的手法呈现出来的作品。而组合的展品，虽然因为尺寸一般比较大而不能参评，但它的题材之丰富，仿如置身于大自然当中，让人们融入到盆上之景，这也是山水盆景的特色之一。

此次大会除了展区以外，还增加了销售区，这是以往展览中少有的。盆景不仅仅是一种艺术，它还是一种物品。20世纪六七十年代的中国，很多老百姓在院子里、天台上都有种盆景，当时虽然资讯没有现在发达，但人与人之间对盆景的交流一点不比现在少，人们用最朴素的手法加上自己的演绎，做出一盆盆各具特色的盆景。然而，要想盆景艺术能够继续发展下去，流通是必然的，除了一些对我们有特殊意义的作品外，其他的作品无论是用作馈赠、交换、买卖，对市场都有正面作用。现今世界，什么都以快优先，盆景能让人们安静下来，欣赏着大自然赋予的生命，通过做盆景，学会规划、学会参与、学会交流、学会欣赏。

很多行外人都担心盆景买回去之后要如何护理，本人希望

Talk About China Ding 2013(Guzhen)

China National Penjing Exhibition

在盆景展览中，能设置一些由参展人自行布展的区域，他们可以将其设计成家居或庭院，以示范单位的形式将盆景融入生活，也能够更好的与参观者进行互动沟通。另外，还可以将第一次参展的作品集中起来，建立新品展区，评选新品奖，以此来激励盆景人做出更多更新的作品。

说中国盆景年度晚宴耳目一新一点也不为过，晚会由多个环节组成，无论是在视觉上还是在听觉上都令人得到了饱足。过往的宴会，一般都是每个协会或每个地区的人坐在一起，南北、各省市地区之间起不到交流。而这次的晚会，格调之高，在全球的盆景晚会中前所未有的。让我们知道，盆景可以是握在手掌之中，也可以是单独成为一景；可以是家居修饰，也可以登大雅之堂；可以是平常人家互相馈赠，也可以拍卖收藏。晚宴的每个环节都有它的特殊含义，个人特别欣赏打鼓的环节，震撼的感觉令人振奋不已；而沙画，却让人们在短短的几分钟之内了解到盆景在中国的发展过程，同时表演者优美的手法，也令人神往。

早在2013年5月份的《中国盆景赏石》已经首次刊登了这次大展的资讯，6月份还出了通告，为这次大展作了详细介绍，让有意参展的盆景人可以提早做准备。那时在广东正值盛夏，考察小组人员每天马不停蹄地实地考察，甄选符合资格的作品上报给评比委员会。看到他们的汗水和疲累，也感受到本届大会展前付出的努力，但更多的是看到他们的耐心和脸上的喜悦！

作为送展人之一，在会场我看到组委会有条不紊的安排。每一位送展人员先到报名处报名，领了票号就耐心等候，组委会是按照次序安排卸车，然后由专人把展品送到它的展位上，整个次序非常清晰。因为这次展览是在室内举办的，因此能让参观者更加舒适的欣赏到每盆作品，很多人第一时间将拍到的照片放在互联网与人分享，也起到了很好的宣传效果。展览现场某些时段在灯光方面还有待改进，以便让拍摄的效果更加完美。如果能够在展场播放一些送展、布展的花絮就更好了，当我们在展场慢慢欣赏如此优雅的盆景作品的时候，我们很难想象它们有些是经过一千多千米、有些是用一百吨的吊车才能卸车，当中有些盆碎了、树折了、配件坏了，但盆景人不是埋怨，不是叹息，而是第一时间看如何能够把它修复，这就是盆景人的修养和品格。

本次2013（古镇）中国盆景国家大展由9位评委和监委组成评委团队，我个人认为这是目前最好的评比方式，记得2012年的中国盆景精品展，在现场和《中国盆景赏石·2013（古镇）中国盆景国家大展全景报道特别专辑》都有刊登评比计分表，非常清晰，我相信参与者大都会认同这种方式的。这次大展非常着重展品的质量，我相信出来的分数会很接近，是否能够获奖跟展品的状态有很大关系。

作为一个盆景新人，体会到盆景艺术在中国乃至世界将会进入一个前所未有的新天地，交流、学习会更加紧密。未来，盆景不会局限于单一的品种和手法，将会有更多新的尝试呈现在人们面前，整体上也会力求完美，让人无可挑剔。中国的盆景人要让人们能看得懂盆景、做得出盆景、懂得如何养盆景。

一个偶然的机会与我的师叔叶国良先生结缘，并重新接触盆景（在我小时候家父也是盆景人），我们的作品有别于传统盆景，一般都比较大，注重组合以及山水人物景观，每一盆盆景都可以独立成景，在不断的摸索当中，得到了很多前辈的宝贵意见和建议，做出了多方面的改善，作品也日趋成熟，路虽漫长，但仍继续！

View on the Differences Between China Ding

– 2013 (Guzhen) China National Penjing Exhibition and Huafeng Bonsai Exhibition

初窥中国鼎

——2013（古镇）中国盆景国家大展和华风展之浅见

文：罗瑞本 Author: Luo Ruiben

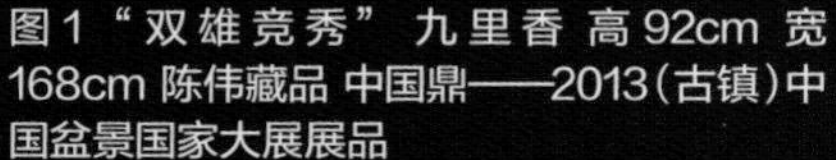
图1“双雄竞秀” 九里香 高92cm 宽168cm 陈伟藏品 中国鼎——2013（古镇）中国盆景国家大展展品

图2 台湾真柏 高75cm 宽170cm 吴明选藏品 中国鼎——2013（古镇）中国盆景国家大展展品

图3 “似狂继癫” 红花檵木 高150cm 吴成发藏品 中国鼎——2013（古镇）中国盆景国家大展展品

图4 相思 高120cm 宽110cm 陈光明藏品 中国鼎——2013（古镇）中国盆景国家大展展品

2013年初次到大陆观赏盆景展，当欣赏完“中国鼎”盆景展之后，脑海中经常浮现着应将中国大陆的盆景特色与中国台湾盆栽之优点做一分析与区隔，期盼两岸能相互欣赏彼此之优点，更进一步相互融合（取彼之长补己之短），创造出更辉煌的成绩，让举世喜爱盆景（盆栽）之雅士，能耳目一新看到前所未有的盆景艺术。

中国大陆地广物博，树种繁多，野采树桩不胜其数，且优秀树桩众多。再加上爱好者长时间培育养成之盆景在质感上常让人觉得古朴生辉，年代久远（如图1~4），使人内心产生一种对老态龙钟或是对年迈长者的敬畏之心，能令收藏家至诚主动地想邀请此类作品回去供养，以利在家方便怡情养性，自娱忘我，时时刻刻能陶醉在艺术生活的美感中。

作者简介

罗瑞本，曾为中华榕树盆栽协会常务理事，中华盆栽作家协会筹备委员。现任嘉义市盆栽艺术协会荣誉理事长，中华盆栽作家协会委员，嘉义县盆栽艺术协会常务理事，中华榕树盆栽协会理事， 嘉义市城隍庙盆栽班讲师。

图 5 华风展展品

图 6 华风展展品

图 7 华风展展品

图 8 华风展展品

中国台湾土地虽狭小，但由于属于亚热带地区海岛型气候，长年受高温多雨多台风的侵袭，且中央山脉高 3952m，所以树种繁多，生长特快，但病虫害种类与危害也特高，因此造就中国台湾农业科技特别发达，盆栽树种的实生栽培技术也就独步全球，无论松柏类或杂木类（常绿类、花果类、落叶类）等树种培育方法与整枝造型技术均可领先世界各地，所以目前中国台湾的盆栽发展及落叶盆栽更能彰显此特色（如图5～8）。

图 9"秋林野趣" 雀梅 高 110cm 陈满田藏品 中国鼎——2013（古镇）中国盆景国家大展展品

中国大陆盆景在创作的技法上以岭南派为中轴（崇尚自然、截干蓄枝）再融合各地方之特色，因此在作品上线条特别生动活泼有力，律动感适当，转弯点与大小长短之比例也掌握得宜。空间的处理也以气韵生动与空灵、寂静之思想为展现之主轴，所以常呈现出大空间来显露出作品骨架的线条美与和谐的比例及古朴苍劲的质感，以突显中国文化之深邃意境（如图9～12）。

图 13 华风展展品

图 14 华风展展品

图 10 “如意”九里香 高 66cm 宽 97cm 徐伟华藏品
中国鼎——2013（古镇）中国盆景国家大展展品

图 11 “三口之乐沐春风” 榆树 高 110cm 长 128cm 陈再米藏品 中国鼎——2013（古镇）中国盆景国家大展展品

图 12 “榕韵” 榕树 高 120cm 洪容兴藏品
中国鼎——2013（古镇）中国盆景国家大展展品

中国台湾盆栽创作的技术在张武德老师的带领下成功地将其盆栽鉴赏与原则的技法推广至今，广受好评并深受盆栽界人士认定与应用，此法严格地要求根、干、枝的出处，线条流向与比例等。再加上台湾盆栽人对树种严格地筛选，选出特优的品种加以培育，又拜地理位置之赐，形成台湾盆栽在管理上，叶团密麻扎实，枝梢纤细绵绵，层次分明，出枝顺畅及八方根盘，使树屹立于盆钵而不动摇，展现平稳端庄的大树相。（如图13～16）

图 15 华风展展品

View on the Differences Between China Ding

-- 2013 (Guzhen) China National Penjing Exhibition and Huafeng Bonsai Exhibition

图 17 “峡江帆影” 真柏、龟纹石 高 60cm 宽 150cm 芮新华藏品 中国鼎——2013（古镇）中国盆景国家大展展品

在文化的表现上，中国大陆盆景以绘画诗词、书法，作为形态表现的主轴，更以儒家重视传神的表达（中庸：“道也者，不可须臾离也”以道为中心的思想。佛家《六祖坛经》云：“自心是佛，更莫狐疑，外无一物而能建立，皆是本心生万种法故。”经云：“心生种种法，心灭种种法”，“吾人顿修自识本心，自见本性，无念法者见一切法，不着一切法，遍一切法，不着一切处，变一切处……我此法者，从上己来顿渐接力于无念为宗。”道宗老子：“有无相生，难易相成，长短相形，高下相倾，复归于朴，大巧若拙。”）等哲理为内涵，展现出作品的幽远意境，再加上野采树桩的野趣天成之优势，所以作品的造型较偏爱生动活泼及空间留白的和谐律动与线条的舞动飞扬如此刚柔并济、疾徐有度的节奏美，除此之外，更讲究叶团疏密关系，与争让关系及宾主关系的处理，从盆面的装饰大巧若拙，即可看出用心之处，其中更以水旱盆景、山水盆景以师法自然景观或师法山水画为模板，再加上作者的巧思去创作，这类的作品在世界各国的展场上是无可见到的，尤其是作品的比例与布局浑然天成给人感受到大自然的野趣及中国文化深远幽长的意境无限，令人回味再三、动荡回肠留下特殊深刻的印象（如图17~20）。有九成以上的作品，均有题名，可见作者或收藏者皆胸存墨宝，且部分题名已达点景引意、景名相符之效果，可引人入胜，使人从内心由意会到景深而意远，寓意于景，情景交融、意犹未尽之感，以此形显露中国盆景之内涵与特殊之处，深深令欣赏者惊艳。

图 18 “山水清音” 水旱盆景 高 120cm 宽 62cm 曹克亭藏品 中国鼎——2013（古镇）中国盆景国家大展展品

图 19 “情怀江绪”水旱相思 高 70cm 何长洪藏品 中国鼎——2013（古镇）中国盆景国家大展展品

图 16 华风展展品

图 20 “天宫胜景”清香木附石 高 130cm 陈迎凯、何兆良藏品 中国鼎——2013（古镇）中国盆景国家大展展品

中国台湾在文化上虽与大陆同根源，但中国台湾的盆栽信息在20世纪60年代后即受日本盆栽观念与技法的影响至今仍存在，虽是如此但中国台湾业者仍以本土树种及气候开发出自己的作法与观点，如从根盘、塑干、整枝与整体造型及枝梢等与叶团之管理上均有精辟细腻的技法与见解。再加上中国台湾市场小、竞争激烈、农业科技进步神速，盆栽界人士慎选出枝叶优良品种，进行实生繁殖，于是创作出现今细叶紧实、枝梢绵密之精巧造型，至于尺寸之大小也订于100cm以内，如此在置场上可方便搬运，在摆设上更利于室内厅房的布置，取轻巧方便之利。在作品内涵上的表达则以植物生理与树种特性为主，重于形、唯妙唯肖、巧夺天工的特点，尽极人工管理之本色（如图21～25）。

在学术的发展方面，目前中国盆景有三大传媒(《中国盆景赏石》、《花木盆景》与《中国花卉盆景》)均有众多专业人士与学者发表各自的研究报告如造型、技法、理论与中国文化内涵之意境及神韵等哲理（如苏本一、赵庆泉、胡乐国、潘仲连、王选民、徐昊、吴成发、刘传刚、谢克英、马文其、吴培德、柯成昆等），及喜爱盆景艺术者均有发表或编著成书可谓研究风气如雨后春笋、百家争鸣、盛况空前般的热络讨论和研究。提供了盆景艺术发展方向的掌舵手，推动了盆景美学、盆景文化的跳跃进步。

图21 华风展展品

图22 华风展展品

图23 华风展展品

图24 华风展展品

图25 华风展展品

图 26“奇劲唱风” 赤松 高 140cm 宽 130cm 曹志振藏品 中国鼎——2013（古镇）中国盆景国家大展展品

图 27 “谦谦君子”山松 高 100cm 彭盛材藏品 中国鼎——2013（古镇）中国盆景国家大展展品

图 28 “礼”真柏 高 115cm 宽 75cm 吴国庆藏品 中国鼎——2013（古镇）中国盆景国家大展展品

图 29 九里香 高 120cm 宽 90cm 肖庚武藏品 中国鼎——2013（古镇）中国盆景国家大展展品

在中国台湾方面则在1970~1980年间有盆景界季刊杂志，由黄德章先生创办，当时也有一群盆景前辈相继投稿发表，其内容大部分是农业栽培的管理知识或经验及整枝造型之技法，与树种特性等方面的探讨与研究，此时也正值中国台湾经济起飞的年代，盆栽市场也随之热络（如同现今中国盆景之盛况一般），想学习盆栽管理或创作者及专业中介买卖者也如同花好招蝶般引来群蝶飞舞，因此孕育而生的盆栽教室相继成立，以传授技法与盆栽管理。也因此，在创作上均较严谨、无论根盘、树干、树枝及枝梢均力求比例和谐、线条顺畅、造型完美，更以黄金比例的三角形为依据，创作出端庄稳重的大树相，在栽培管理上更深入植物学、介质与肥料学、植物保护学等学科作为盆栽用土（介质）的组合、换盆、摘芽、疏叶、疏花、疏果、剪枝、造型、扦插、嫁接、施肥、移植等季节的时机性与病虫害防治进行周期化的管理，可谓对盆栽管理与技法的研究用心至极，此乃中国台湾盆栽之强项也。

中国大陆盆景喜爱气势磅礴、苍古雄劲、雄浑厚重的气势，以突显泱泱大国之姿（如图 26 ~ 29），但此等意境的传达无关盆景尺寸之大小，应依各民族审美意识与风俗民情之普遍性做弹性择优完善而定行之。除此之外，在主观的意识上，更以儒家、道家、佛家之哲学思想做为创作内涵，再融入山水画及诗词之意境，使作品在世界舞台上展现出与众不同的风格，并且已开始对世界各国产生影响力，此乃盆景起源国之软实力的展现。在中国台湾方面显然在农业栽培管理与造型技法上较严谨精致。总而言之，中国大陆盆景在质感、空间与意境的表达较优秀，在人文上面，文学之涵养与修为也普遍具备高水平，而中国台湾盆栽在枝梢、叶团与树种特性美较见长，在农业科技的研究与管理也较完善。倘若两岸盆景界人士能持续深入交流，融合彼此之长处相信不久便能执世界之牛耳。再假若今后的创作者能更精勤进取将东西美学融入作者心中，再经由各自内在的知识，生活经验与文化涵养等哲学意识形态，重新再塑造一个具有自己性格理想的自我，（切记不可损及自己之优点与特色，否则画虎不成反类犬）届时作品将会是带领盆景艺术走进辉煌灿烂的新里程碑，开创中国盆景的新一页，吾在此预祝盆景艺术未来闪亮光芒的大道，能够是两岸盆景雅士所铺设的。

嘉宾团在唐苑合影留念 纪武军摄影

2013 Chin

2013中国鼎之旅

——2013（古镇）中国盆景国家大展后中外嘉宾参观团访问之中国唐苑篇

本文图片除署名外，均为本书编辑拍摄

唐苑鸟瞰图 刘伟民供图

Foreign Guests' Delegation Visited Chinese Tangyuan Garden after 2013 (Guzhen) China National Penjing Exhibition

2013 年 10 月 2 日，陕西万达实业有限公司董事长、世界盆景石文化协会会长张小斌，中国盆景艺术家协会会长、世界盆景石文化协会名誉会长兼首席全权代表苏放，中国盆景艺术家协会副会长、世界盆景石文化协会秘书长樊顺利，陕西唐苑园林观光有限公司总经理韩继明等陪同 2013（古镇）中国盆景国家大展中外嘉宾游览中国唐苑。

中国盆景艺术家协会副会长樊顺利与小林国雄探讨盆景的养护

2013 China Ding Trip

Foreign Guests' Delegation Visited Chinese Tangyuan

Garden after 2013 (Guzhen) China National Penjing Exhibition

唐朝，曾是中国富甲天下傲视世界的时代，以"唐"为名的唐苑是凝聚了建立者多年来内心愿望的一个中国盆景人的文化寄托，参观完中国盆景国家大展之后的中外嘉宾团参观的最后一站就是《中国盆景赏石》的很多编辑也从未到过的西安唐苑。这里的参观，引起了所有国际嘉宾团成员的不绝赞叹！在这里，他们再一次感受到：当代的中国盆景私人园林在中国强大的经济发展的支撑下正在从世界盆景园林舞台上冉冉升起并崭露新的光芒。

2013年10月2日，2013（古镇）中国盆景国家大展中外嘉宾参观团在陕西万达实业有限公司董事长、世界盆景石文化协会会长张小斌，中国盆景艺术家协会会长、世界盆景石文化协会名誉会长兼首席全权代表苏放，中国盆景艺术家协会副会长、世界盆景石文化协会秘书长樊顺利，陕西唐苑园林观光有限公司总经理韩继明等的陪同下参观了这座位于古城西安东南部汉代苑林遗址杜陵塬上的名震中国的唐苑。

中国唐苑，这个因"唐风"盆景展等盆景活动和事件而享誉海内外的私人出资建造的盆景公园，它集中国传统文化与生态旅游为一体，占地约$1.3\times10^6m^2$（2000亩），其设计主题为"让辛勤的人们回归大自然"。从唐苑的平面设计图上可以看到唐苑包括观光旅游区、园林艺术区、餐饮游乐区、博览会馆区、健康管理区、休闲度假区、温泉

丰腴的根系 刘伟民摄影

百米瀑布 刘伟民摄影

小林国雄与须藤雨伯就生病针叶进行研究 纪武军摄影

盆景园一角 刘伟民摄影

洗浴区、高尔夫球场区、汉唐文化区9个区域，是真正的集旅游与休闲、健康与文化于一体的艺术与生活的交汇地。进入唐苑，满目皆绿，垂柳、古槐、黑松、银杏、皂角等争奇斗艳，给人一种返璞归真之意境。唐苑收集了不计其数的关中民俗石器。由石碾石磨铺就的观光道路旁有一百只形态各异的石羊，寓意着“百羊开泰”，另一侧则是由马槽做成的荷塘，种植着各种荷花、睡莲、水仙，每件民俗石器都烙印着岁月的痕迹，铭刻着祖辈们对美好生活的向往与追求。如果说巨石、奇石是唐苑的一大特色的话，那么百米瀑布便是唐苑一道亮丽的风景线。气势磅礴的百米瀑布与其边上纵横生长的苍劲黑松构成了一幅烟雾缭绕的人间仙境，吸引众嘉宾驻足拍照。飞流直下的瀑布汇入潺潺溪流中，溪水中的日本金鲤鱼成群结队，俯瞰水面，一群群红黄色的鲤鱼穿梭于绿树的倒影中，动静结合，相

小林国雄，须藤雨伯，梁悦美等专家在探讨生病盆景的病因 纪武军摄影

得益彰，来自世界各国的嘉宾们都激动不已地按下快门开始记录唐苑里的每一个美丽的场景，穿过蜿蜒的折叠木桥，呈现于眼前的是铭源祥红木坊工艺制作车间及成品展区，走进满是木香的制作车间，看到工艺师傅们精湛的制作技艺，嘉宾们不禁竖起了大拇指。

唐苑聚集了两万余盆不同风格不同品种的盆景和素材。古朴沧桑的榆树、隽秀挺拔的罗汉松、争奇斗艳的杜鹃、枝繁叶茂的白杨、千姿百态的紫薇，还有苍劲高大的黑松。来到唐苑就不得不看唐苑的黑松园，这里的黑松无论是造型还是养护都

秋意浓浓 刘伟民摄影

嘉宾们体验唐苑的观光车

参观团抵达唐苑国际会展中心

水畔形态各异的黑松向来是游客们镜头中的宠儿 纪武军摄影

造型别致的石雕 纪武军摄影

折桥边古色古香的建筑物 纪武军摄影

让人眼前一亮。说到盆景，就必不可少的要想到盆景的养护，这次参观也正是国内外专家们探讨黑松养护的一个绝佳机会。在黑松园里，中国盆景艺术大师樊顺利大师指出两株针叶干枯的黑松，虚心向各位国外专家请教，日本盆景专家小林国雄还亲自爬到高大的黑松顶部去采集样本，并分发给各位专家。专家们拿到样品后，通过观察其颜色、闻其气味、品尝其叶子的味道等来判断得病原因，最终小林国雄和须藤雨伯表示回国后会寄药剂到唐苑。走出黑松园，天色已经接近黄昏，嘉宾们意犹未尽的表情是对唐苑的高度赞赏，也是对此次游览活动的赞许和肯定。

中国唐苑——这片曾经举办过"唐风"盆景展、"唐风"赏石展、"唐风"兰花展、红木家具展、海外文物展、四季花卉展、汉唐文化展等展览的土地上迎来了世界各地17个不同国家和地区的盆景专家和杂志媒体，就各国的盆景历史、现状和未来

水边小景 纪武军摄影

2013 China Ding Trip

Foreign Guests' Delegation Visited Chinese Tangyuan Garden after 2013 (Guzhen) China National Penjing Exhibition

国外嘉宾欣赏盆景 纪武军摄影

中国盆景艺术家协会会长苏放与国外嘉宾交流盆景

陕西唐苑园林观光有限公司总经理韩继明给参观团讲解红木家具

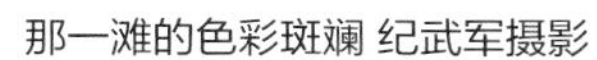

那一滩的色彩斑斓 纪武军摄影

由磨盘组成的特色小道 刘伟民摄影

精品苑一角 刘伟民摄影

2013 China Din

Foreign Guests' Delegati Visited Chine

Garden after 2013 (Guzhen) China National Penjing Exhi

极具中国特色的雕像吸引众嘉宾纷纷举起了相机 纪武军摄影

中国盆景艺术家协会会长苏放与波兰盆景协会会长沃齐米日 · 皮特思科（Wlodzimierz PIETRASZKO）交流盆景知识

具有千年树龄的蔷薇

小林国雄爬到树上采摘样本 纪武军摄影

粗壮的根盘 令人震撼 刘伟民摄影

发展进行了深入的探讨和交流。2013“唐苑的世界盆景对话”国际年度论坛于2013年10月3至4日在唐苑酒店1号会议室召开，此次论坛的详情内容会在下月的《中国盆景赏石》中与大家见面！国外著名盆景人在参观“唐苑”、参与国际论坛后亦是赞不绝口，在下月的专辑中，读者也将看到国外盆景人眼中的“唐苑”及国际论坛！精彩还在继续，敬请期待！

祈福殿 刘伟民摄影

镜头下的百米瀑布 纪武军摄影

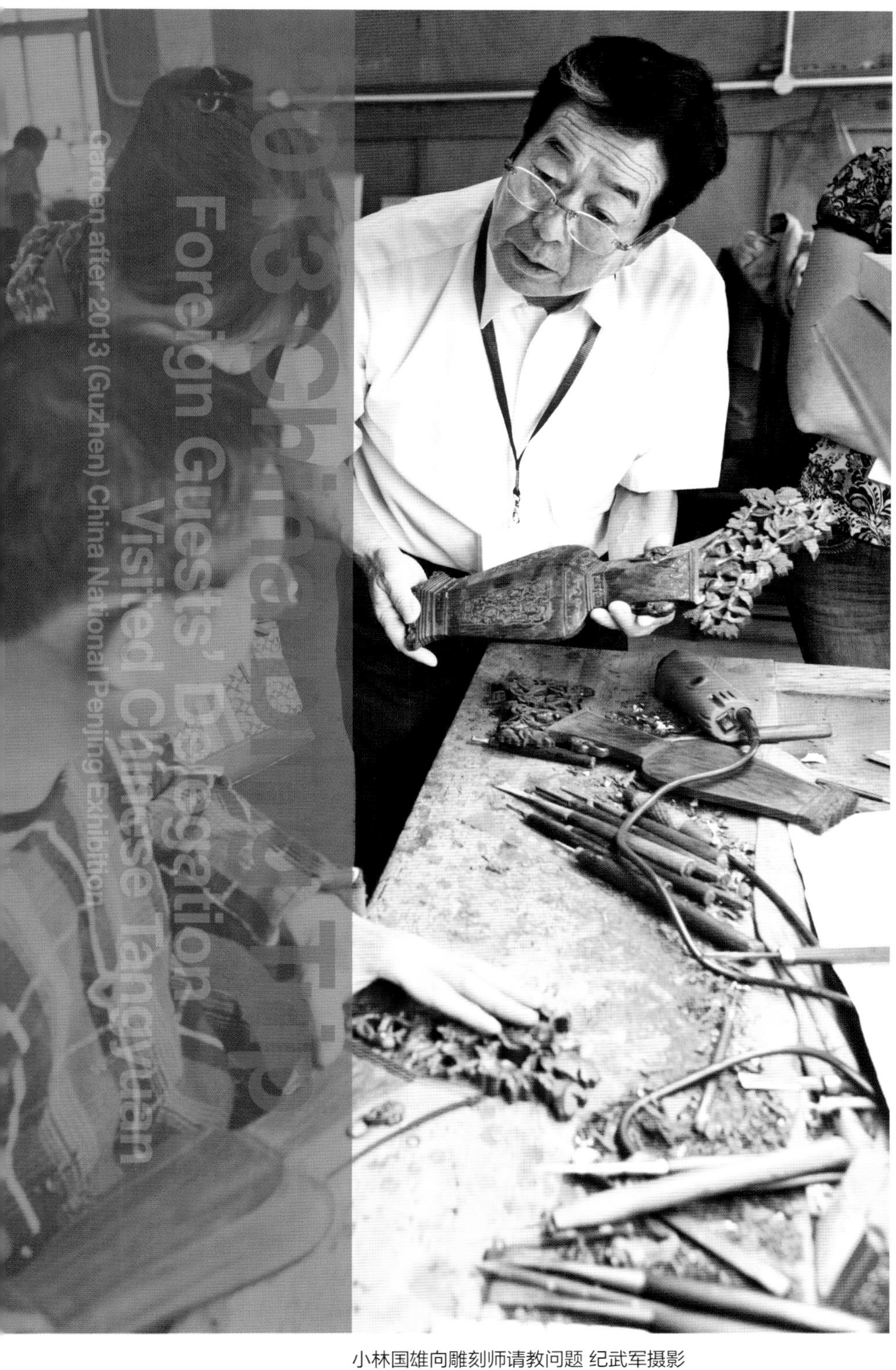
Foreign Guests' Delegation
Visited Chinese Tangyuan
Garden after 2013 (Guzhen) China National Penjing Exhibition

小林国雄向雕刻师请教问题 纪武军摄影

苍劲的树干

这棵树的直径是多少呢 刘伟民摄影

“唐苑之王”刘伟民摄影

活动时间：2013 年 9 月 30 日~ 2013 年 10 月 1 日
活动地点：广东省深圳市罗湖区东湖水库旁趣怡园

2013中国鼎之旅

——2013（古镇）中国盆景国家大展后中外嘉宾参观访问之趣怡园篇

2013 China Ding Trip

— Sino - Foreign Guests Visited the Quyi Garden after 2013 (Gu Zhen) China National Penjing Exhibition

参观团在趣怡园合影

趣怡园赏石展区一瞥

美国《国际盆栽》杂志出版人兼主编 威廉·尼古拉斯·瓦拉瓦尼斯正在拍摄喜欢的盆景

2013年9月30日即2013(古镇)中国盆景国家大展后，为了促进中外盆景文化交流，中国盆景艺术家协会组织几十人的中外嘉宾团在中国盆景艺术家协会副秘书长邓孔佳的引领下，于次日清晨前往趣怡园参观访问，通过喧闹的深圳市中心，到达罗湖区东湖水库附近的一个宁静的私家盆景园林——趣怡园。

参观访问团欢迎晚宴

9月30日抵达当晚，趣怡园园主吴成发先生不仅为远道而来的国内外宾客安排入住深圳五星级

日本水石协会理事长小林国雄先生（左一）和景道二世家元须藤雨伯先生（右二）探讨中国古盆

树下的人物配饰使得画面栩栩如生

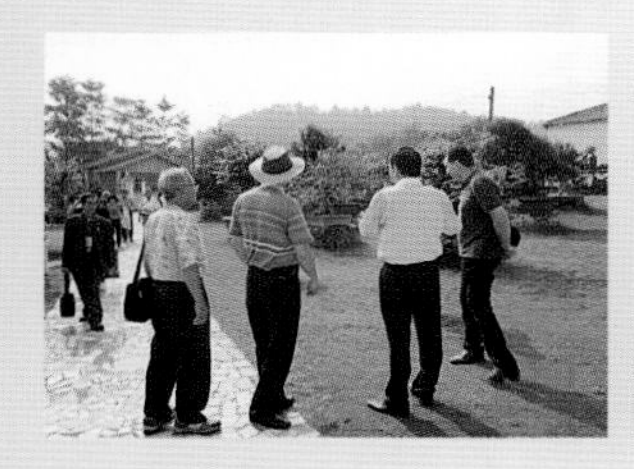

园主吴成发先生（右二）为中外嘉宾讲解趣怡园和中国盆景

欧洲盆栽协会成员国瑞典盆景协会理事会成员 玛丽亚·阿尔博尔莉思·罗斯伯格女士正在观看中国名石

美国《国际盆栽》杂志出版人兼主编 威廉·尼古拉斯·瓦拉瓦尼斯正在拍摄中国独具特色的石头

国际盆栽俱乐部意大利理事 玛利亚·基亚拉·帕德里奇女士正在研究中国奇特的石头

这棵树的做法深受嘉宾们的喜爱

标准豪华酒店，而且为来宾举行热情洋溢的欢迎晚宴，晚宴上准备了丰富多彩的节目，除祝酒致辞外，有以抽奖形式赠送名家名画的趣味活动，这些名家名画不仅向外国友人展示中国又一宝贵的传统文化而且给予国内书画爱好者视觉享受。还有欢乐颂歌和集体热舞等节目，嘉宾们有的唱歌有的随着动听的音乐翩翩起舞，晚宴欢乐的气氛将此次参观访问活动推向高潮。

趣怡园概况

趣怡园建于20世纪末，占地30多亩，依山傍水，景色宜人。趣怡园是由

园内风景

古盆赏石收藏长廊

趣怡园中古盆展区一角

精美的盆景吸引来宾拍照

邓孔佳先生（左一）带领中外嘉宾参观趣怡园中的古盆赏石

美丽的盆景吸引法国《气韵盆栽》主编米歇尔·卡尔比昂上前拍照

中国盆景艺术大师、世界盆景友好联盟（WBFF）国际顾问吴成发历经近20年艰辛打造而成的岭南盆景创作和展示基地，不断推动中国盆景尤其是中国岭南派盆景走向世界。同时此处也是“世界盆景友好联盟国际交流中心”和“香港盆景雅石学会（深圳分会）”所在地。园区科学的规划和合理的布局得到了国内外的一致好评和高度认可。这里创作了高质量、品种多的各式盆景3000多盆（很多是获奖作品），以及明、清、民国盆景古盆3000多个，是享誉业界的顶级盆景园，2012年9月，“趣怡园”中6盆大型盆景被意大利米兰克雷斯皮盆景博物馆永久收藏，为中国盆景走向世界做出了贡

美丽的趣怡园园林入口

园内荷花盆

献。作为专业中国岭南派私家盆景园，其规模之大、质量之高、技术之精、养护之好，堪称“第一中国盆景私家园林”。

广东趣怡园推广中国盆景走向世界，打造岭南派盆景文化交流教育基地

园区主要特点：立体盆景园林展中华，超多古盆、雅石移步景！为何有如此的惊叹，缘由如下：第一，与大多国内私家盆景园不同的是，趣怡园是纯粹的盆景园，园里不设住宅建筑、假山等，以免打扰其幽静而美丽的景象，而园中的主角只有一个典型的中国岭南派的精华集萃——中国的杂木类盆景。盆景排列得错落有致且空间充裕，园艺工人们细致入微的修剪和精心的养护使之长势茂盛，绿绿葱葱。第二，透过杂木类盆景独有的视角，趣怡园淋漓尽致地展现了中国盆景作为传统文化之一独立和完整的遗传特征，并以

“似狂继癫”
红花檵木 高 150cm

“苍龙奔江”
九里香 飘长 115cm

“王者之尊”
香兰 高 110cm

清中期 紫泥绳纹边腰线圆形盆

明中期白泥清釉兽头波纹口圆形盆

“清雄雅健”盆景

趣味横生的赏石与古盆

趣怡园收藏的精美赏石

趣怡园收藏的精美赏石

趣怡园收藏的精美赏石

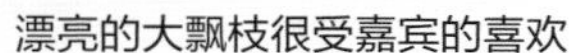

漂亮的大飘枝很受嘉宾的喜欢

《近代盆栽》月刊主编镰田照士先生（左一）、日本水石协会理事长小林国雄先生（右中）和景道二世家元二世须藤雨伯先生（右一）合影

嘉宾在园中休息区拍摄全景照

继承中国传统文化实体的方式向世人展示不同的中国盆景的哲学理念。第三，虽说园中盆景之惊艳，但红花还需绿叶配，如果说盆景是主角，那么蕴含了哲学和工艺技术于一体的很高深的特殊工艺烧制出来的专用古盆和设计优雅别致的几架就是园中的绿叶，彼此相互衬托，相得益彰。

趣怡园园主吴成发先生为来宾精心准备的签到薄

亮点：雅石古盆齐聚一堂

园区左侧通过一道长廊便是广阔的中国岭南派盆景园主展区，参观气氛踊跃，宾客们有的合影留念有的向园艺师傅请教、探讨岭南盆景独有的“蓄枝截干”技艺。期间园主吴成发先生认真而专注地向来宾用双语讲解趣怡园中由于这个特殊技艺而呈现出来的“一树一样”的多样化的空间结构特点的盆景。来自马来西亚的蔡国华先生激动地说：“我非常感谢吴成发先生的热情接待，趣怡园里收藏了非常优质的山采珍贵盆景树种，不仅开阔了眼界还让我了解到岭南派盆景的技法以及具有个人风格的创造力和表现形式的盆景美学。”来自美国的威廉·尼古拉斯·瓦拉瓦尼斯先生高兴地说：“趣怡园不仅展示有不同规格形状的各式各样的、品种繁多的盆景，而且可以看到如此多的中国古盆，令我震惊！一些著名的古盆我以前只在日本见过，这次在中国深圳的一个私家盆景园中也能见到令我惊喜！”

中外嘉宾们与园主吴成发先生（中间左一）合影

日本水石协会理事长小林国雄先生对中国古盆产生浓厚的兴趣

日本水石协会理事长小林国雄先生（左一）仔细端详古盆

临行前，为了激励广大盆景爱好者的创作热情，共同推动盆景文化（产业）的发展，吴先生为前来参观的宾客每人准备一份精美的纪念品，除了一些小巧实用的礼物外，还赠送了《吴成发盆景艺术》一精装套书，其中包括《吴成发先生古盆藏品集》和《吴成发大师盆景作品集》。随后大家依依不舍地与园主及趣怡园的所有工作人员一一道别，感谢之情不尽言表，由衷地说广东深圳趣怡园真正推动了中国盆景尤其是中国岭南派盆景的发展，努力创造中国盆景与世界盆景文化交流平台。

在趣怡园中留下的美好记忆

【意大利】 玛利亚基亚拉·帕德里奇 BCI（国际盆景俱乐部）理事

[Italy] Maria Chiara Padrini BCI Director

在中国的后几天行程中，我们参观了两个私家盆景园，我十分欣赏位于深圳的趣怡园和位于东莞的真趣园。这两个园子让我了解和欣赏到不同形态和种类的树。非常有趣的是看到盆景融入室外空间和与周围环境自然和谐的共处。私人紫砂古盆和赏石收藏令人印象深刻，还有两位园主对我们的热情款待。

I really appreciated the visits to the Quyi Garden in Shenzhen and Zhenqu Garden in Dongguan. Both have allowed me to know and appreciate different visions of Penjing and the variety of trees. Very interesting to see how the Penjing is integrated in wide open spaces and harmonized with the environment. Impressive private collections of antique vases and stones and exquisite hospitality offered to all of us.

【日本】山田登美男 日本盆栽作家协会会长

[Japan] Tomio Yamada Chief of Nippon Bonsai Sakka Association

趣怡园和真趣园都特别热情地款待了我，我感到中国的私人庭院满载希望、前途光明。祝愿今后它们会有更进一步的发展！

Quyi Garden and Zhenqu Garden have hospitably entertained me. I feel that the private gardens of China are full of hope and have bright future. Best wishes for them to have further development!

「日本」山田登美男　日本盆栽作家協会会長

趣怡園·真趣園、共に大変お世話になりました。中国の私人庭園は希望に満ちて明るく感じた。今後、一層発展されることを願っている。

【越南】 阮氏皇 越南盆景协会主席

[Vietnam] Nguyen Thi Hoang President of Vietnam Bonsai Association

吴成发先生的趣怡园设计的非常好，花园中的每个盆景都得到了精心的照料。花园中的树木都具有不同的类型和风格，适用于那些具备盆景知识的参观者。这真的是一个能展示特定盆景风格和中国盆景审美概念的好地方。

About Quyi Garden: It is well designed and each Penjing in the garden is taken great care of. The trees in the garden are diverse in kinds and style. It is suitable for visitors who really have knowledge about Penjing. It is a good place to display the specific Penjing style and bonsai aesthetic concepts of China.

The Wonderful Memory of Quyi Garden

【韩国】 成范永 世界盆栽友好联盟顾问 思索之苑苑主

[Korea] Sung Bum-young
Advisor of World Bonsai Friendship Federation Owner of Spirited Garden

我曾先后两次访问深圳的趣怡园。这里所展现的盆景艺术独具特色，好作品非常之多，珍贵的古盆理所当然是世界上最多的，是一座非常了不起的私家盆景园林。他们拥有雄厚的实力，希望今后能建设一座漂亮的、大规模的博物馆。

I've been to visit Quyi garden two times in Shenzhen. There are many good works to show Chinese Penjing art, the precious ancient pots are deservedly the most in the world. It's a wonderful place. They have a solid strength, but I hope they would build a beautiful and massive museum in the future.

【马来西亚】 蔡国华 国际盆栽俱乐部会员 马来西亚盆景雅石协会会员

[Malaysia] Dato'Chua Kok Hwa
Member of Bonsai Clubs International, Member of Malaysia Bonsai N Stone Association

我必须感谢吴成发先生，感谢他的款待和晚宴上丰盛的美食。吴成发先生收藏的盆景都是很好的树种，这些都是山采的非常珍贵的收藏品。他的岭南派技法、个人风格、创造力造型和盆景美学都是值得我们学习的。

I must thank Mr. Ng Shingfat for his hospitality and warm reception at the Quyi Garden and dinner party. Mr. Ng.has very good collections of Penjing species collected from the wild which are very precious. His Lingnan techniques, personal style, creativity, form, and aestheticsin Penjing are something we can learn from.

【西班牙】 安东尼奥·帕利亚斯《当代盆栽》杂志执行主编

[Spain] Antonio Payeras
Editor and Director for *Bonsai ACTUAL* Magazine

在中国盆景国家大展后我们参观了两个中国顶级私家盆景园林。在深圳吴成发先生的趣怡园，我看到那么多不同树种、不同类型的盆景，真是一个宏伟壮观的盆景园。

We visited two excellent private Penjing gardens after the National Penjing Exhibition, in Mr. Ng Shingfat's Quyi garden I saw so different Penjing species and styles, all magnificent. A great group of trees that rarely can be seen together. I could appreciate too, one of the most complete antique pot collections, with unique exemplars of some of the best pots in the world.

The Wonderful Memory of Quyi Garden

【印度】苏杰沙 印度盆景大师

[India] Sujay Shah
India Penjing Master

深圳的趣怡园和东莞的真趣园都非常漂亮。坦白地说，这两家私人盆景园林给我的感受差不多！园林布局都很有艺术性，从中我受益匪浅。园林的主人也非常热情好客。

We visited Quyi garden in Shenzhen and Zhenqu garden in Dongguan during the 30th of September and the 1st of October. Both of the gardens were very beautiful. Frankly, I have the same feeling to them! Gardens were made very artistically and we got a lot of ideas from them! The hosts were also very hospitable.

【日本】镰田照士《近代盆栽》月刊主编

[Japan] Terushi Kamada Chief Editor of the monthly magazine *KINBON*

我不能想象趣怡园只是爱好者的展所，超过1m的大型盆景在广阔的场地上一字排开，我想这里恐怕聚集了大部分中国本土的东西吧。在日本从未见过的盆器、和日本水石完全不同的赏石美得出众。

I cannot imagine that Quyi Garden is just an exhibition place for fans. Large-scale Penjings exceeding 1 m are lined up in the broad venue. I think most local artworks in China are gathered here. The pots which I've never seen in Japan and the scholar's rocks which are totally different with Japanese suiseki are remarkably beautiful.

「日本」鎌田照士『月刊　近代盆栽』の編集長

いずれも広大な敷地に1mを超える大きな盆景がずらりと並び、趣味家の棚とは思えない趣怡園では、おそらくほとんどが中国本土で集められたものと思われる、日本では見たことのないような鉢や日本の水石とは違った石の蒐集の凄さにも圧倒された。

【丹麦】汤米·尼尔森 EBA欧洲盆景协会成员国丹麦盆景协会会长

[Denmark] Tommy Nielsen
President of Danish Bonsai Association

在中国游览趣怡园的经历令人印象深刻，也是我最喜欢的事情之一。感谢吴成发先生的热情邀请。他制作的盆景树木非常漂亮。非常高兴在他的私人收藏中看到石头和古代的盆景盆。通过欣赏趣怡园的建筑来感受它的美以及它散发出的自然之感。

很高兴能够有时间到处走动，并享受所有盆景树木带来的无法形容的美感。吴成发先生非常的热情好客，他也愿意为我们介绍趣怡园。在我看来吴成发先生是一位非常了不起的人。同样让我高兴的是，我有幸在前一天晚的抽奖活动中幸运的获得了一幅绘画作品作为奖励。我想我这辈子都不会忘记这次经历。趣怡园中一棵小型的黑松同样也吸引了我的眼球，虽然它看上去很小但是它却很强劲。如果要我说哪棵树木是趣怡园中最好的树木，我一定会说是这一棵。相比其他树木，这也是一生中只能找到一次的树。

Quyi Garden was a very impressive experience, it was one of the things i liked the most first of all because Mr.NG Shing Fat had invited us personally and we had made a relationship with him the night before.His trees were amazing and it was nice to see the stones and ancient pots in his special builder house...It was a garden where you could use houses and houses just to look and feel the please and the nature aside it...It was nice to have time to walk around and just be your self and enjoy the incredible work on all the trees...Also the hospitality of Mr.NG Shing Fat and his willingness to tell about the garden was very nice.

My personal view of Mr.NG Shing Fat was that he was a great man in every way, i was very happy also when i was one of the winners in the lottery at the knight before and the painting will be a very remarkable memory that will last for the rest of my life as for the books he gave to us...

One tree that caught my mind in the garden was a very small black pine, it was very small but very powerful, if i should tell wish tree that was the best in the garden that would be that one, but compared to all the other trees this was also one you only find once in a lifetime...

【法国】克里斯蒂安·弗内罗法国《气韵盆栽》杂志出版人

[France] Christian Fournereau, Chief Publication for Esprit Bonsaï Magazine

【法国】米歇尔·卡尔比昂 法国《气韵盆栽》杂志主编

[France] Michèle Corbihan, Chief Redactor for Esprit Bonsaï Magazine

吴成发先生的趣怡园非常有趣，因为它既是一个私家园林同时也体现一个人的情感。在他看来，他的感情以及他个人的处事方式都能体现他对盆景的热爱。我深感荣幸能够欣赏这些极具个性的树木，并能这么近距离地欣赏它们。非常幸运的是，我们还能欣赏他的古盆收藏，这些收藏在欧洲实属罕见。

The private garden of Mr NG Shing Fat, was especially interesting, because it is the work and the sensibility of one man. It was his point of view, his feelings, his personal way of arranging, presenting his love for Penjing. It was a great honor to be able to admire trees with such a personality, and to be able to appreciate his work closely. We were very lucky to be able to admire his ancient pot collection. These collections are so rare in Europe.

【瑞典】玛利亚·阿尔博尔莉思·罗斯伯格 EBA 成员国瑞典会长

[Sweden] Maria Arborelius-Rosberg President of Swedish Bonsai Association

趣怡园景色宜人，坐落在风景秀丽的地方。我为有这么多美丽的艺术品和古盆而感到震惊。园主吴成发先生热情好客，慷慨大方，为我们提供了精美的礼品和高档的食宿。午饭非常美味，趣怡园为所有客人提供了鲜美的水果，让我们游览之余还能在休息时品尝。晚餐和酒店住宿都非常棒，大家都玩得很开心！

The Quyi garden visit was very nice and had a lovely location. It was impressing with all the beautiful items of art and pots. The lord was very nice and welcoming and very generous with fine gifts and accommodation. The lunch was very good. Nice with all the refreshments served in the garden. The dinner and hotel was very fantastic. It was great fun!

【匈牙利】阿提拉·鲍曼 EBA（欧洲盆景协会）成员国匈牙利盆景协会副会长

[Hungary] Attila Baumann Vice President of Hungarian Bonsai Association

趣怡园有很多高水平的盆景作品，尤其是簕杜鹃盆景给我留下了很深的印象。赏石的收藏显示出园主高雅的品位，各种不同的色彩、规格还有形状完整的展现出大自然的多姿多彩。古盆的收藏仿佛把我们带回过去一般，通过古盆上的图画我们可以对于过去的历史窥探一二。许多形状、色彩和图画……古盆可以启发今人运用这些古老的形状和色彩对制陶手工艺人表达敬意。很遗憾的是，趣怡园和其中所收藏的藏品对于欧洲人来说还很陌生，所以我非常高兴可以有这个机会参观趣怡园。

The Quyi Garden was a big pleasure to visit this garden and the high quality trees. Especially the bougainvillea trees were very impressive for me. The stones I saw shows an excellent taste of the owner, very different colors, sizes and shapes can fully represent the wonderfully nature, which forms the stones. The collection of the ancient pots brought us back in time, and we could have a look via the painting on the pots into those centuries. Many forms, many colors and fascinating paintings … these pots can inspire people to use these old forms and colors to pay tribute to the old potters. Unfortunately this garden and collection was not known in Europe, so I am very happy, that I had the possibility to visit it.

2013中国鼎之旅

——2013（古镇）中国盆景国家大展后中外嘉宾参观访问之真趣园篇

2013 China Ding Trip

— Foreign Guests' Delegation Visited Zhenqu Garden after 2013 (Guzhen) China National Penjing Exhibition

真趣园中盆景座谈会一角

真趣园中盆景座谈会一角

嘉宾在中国盆景艺术家协会副秘书长、真趣园园主之子黎炽雄（右一）的带领下参观

2013 Chin

真趣园中的盆景

嘉宾集体合影

真趣园园主黎德坚为东莞市政府副秘书长张永忠（左）介绍真趣松

中国盆景艺术家协会会长苏放和真趣园园主黎德坚在活动上合影

想必对大家来说真趣园已经不再陌生，无论是山水庭院，还是盆景奇石，亦或是根雕、古董，国内外嘉宾都早有耳闻，而更让大家迫不及待想早点前往一见的是罗汉松中的珍贵品种——真趣松。

真趣松，学名真趣罗汉松，罗汉松科，与松杉柏等同属花不明显的裸子植物门。是真趣园园主黎德坚在长达20年从事海岛罗汉松种植时发现的变异品种，后经近10年孜孜不倦的研究、选种、育种、试种，终于培养出集新颖性、特异性、一致性和稳定性于一身的罗汉松新品种。真趣松叶片比普通海岛罗汉松的短、宽、厚、亮，每年吐芽4次，吐芽时间长达40多天，这些特点使它成为欣赏之佳品、盆景与景观树之优秀素材。而最令人惊奇的是其叶片呈红黄色，每次吐芽就好像雏菊绽放一样。同时，真趣松的成功育种不仅添补了东莞园林育种的空白，更是填补了园林中裸子植物除了少数红色落叶和洒金叶外尚无色叶品种的空白！因此，有人称赞真趣园时说“如花似锦真趣松，真情真景真趣园”！

真趣园中的盆景

a Ding Trip

真趣园山石一景

真趣园中的盆景

真趣园景色

园主黎德坚在注重新品种植物的开发、培育与创新的同时，还秉承创新与传统并存、景物与人文共盛之风。真趣园以“欣赏自然，思索人生，扩大视野，宁静致远”为建园理念。园区景色错落有致，主体格局曲径通幽，夜幕初上，灯笼点起，九曲回肠的小路，绿树与红灯交相辉映，投影在湖中的水波里，尽显中国风情。近年来，园主黎德坚在真趣园举办海岛罗汉松技术交流座谈会、岭南盆景创作研讨会、国家新品种真趣松学术论坛、国际盆景交流会等活动，并以盆景会友，常常接待国内外盆景友人，为推动盆景事业发展、加强国际盆景交流做出贡献。

园主黎德坚为当地领导介绍真趣松

日本《近代盆栽》月刊主编镰田照士（左）和日本盆栽大师小林国雄（中）向园主黎德坚（右）赠书

嘉宾参观真趣园

真趣松橘红色的“花”

真趣松红色的“花”

真趣园景色

真趣园景色

2013年10月1日，2013（古镇）中国盆景国家大展后中国盆景艺术家协会组织的中外嘉宾参观团应邀参观真趣园。嘉宾们一进真趣园就不禁感叹起来，有的急忙拿出相机拍照，有的仔细观察，啧啧赞叹，显然，大家被眼前的真趣松吸引了，红色的针叶在枝头绽放，如花似锦，灿烂夺目，远远望去好似千树万树桃花开，甚是惹人喜爱。夺目而出的那棵树势更是雄壮强劲，如绝崖古松，有俯瞰五岳之势，似群雄之王，这便是树龄在百年之上、树高达3.38m的真趣松王了！

真趣园中的盆景

真趣园中的盆景“粤韵雄风”九里香 盆长200cm

夜幕初上的真趣园

中国盆景艺术家协会副会长、真趣园园主黎德坚致开幕词

中国盆景艺术家协会会长苏放（右）致发言词

美国《国际盆栽》出版人兼主编威廉·尼古拉斯·瓦拉瓦尼斯发表游园感言

广东省东莞市政协主席刘树基致发言词

东莞市政府副秘书长张永忠

中国盆景艺术家协会名誉会长梁悦美发表游园感想

来自印度的盆景大师苏杰沙在喜爱的同时，还希望带真趣松的小苗回印度，让这美丽的“花之树”也在印度的土地上盛开。

参加本次活动的除中国盆景艺术家协会会长苏放带队的中外嘉宾参观团外，东莞市政府副秘书长张永忠、广东省东莞市政协主席刘树基、东莞市林业局局长胡炽海等当地领导也应邀出席。中国盆景艺术家协会副会长、真趣园园主黎德坚致开幕词，东莞市政府副秘书长张永忠、广东省东莞市政协主席刘树基、中国盆景艺术家协会会长苏放依次致发言辞，来自中国台湾的中国盆景艺术家协会名誉会长梁悦美、来自日本的日本盆栽作家协会会长山田登美男以及来自美国的《国际盆栽》出版人兼主编威廉·尼古拉斯·瓦拉瓦尼斯代表中外嘉宾团分别发表了游园感言，对盆景的技艺和感悟互相交流，并对真趣园的招待致以了由衷的谢意。广东电视台及东莞当地多家媒体对本次活动做了详尽的报道。

在盆景这个共同的爱好下，在艺术的世界里，国与国之间没有了界限，不同的语言也可以交谈甚欢。夜幕下的真趣园展现出东方的魅力，乐队的音乐奏响西式的情调，两者却相互协调融合得恰到好处。各国的盆景人都在这里分享着盆景带来的喜悦，伴着欢声笑语的交谈大家享受了一个有中国菜肴、变脸特技和主人以抽奖方式赠礼的愉快之旅。

夜幕初上的真趣园

夜幕初上的真趣园

日本盆栽作家协会会长山田登美男发表游园感想

广东省东莞市政协主席刘树基为嘉宾抽出真趣园的纪念品

真趣园中的盆景 真趣松 高 150cm

夜幕初上的真趣园

嘉宾在真趣园中休息交流

嘉宾参观真趣园

在真趣园中留下的美好记忆

The Wonderful Memory of Zhenqu Garden

【韩国】成范永 世界盆景友好联盟顾问 思索之苑苑主

[Korea] Sung Bum-young Advisor of World Bonsai Friendship Federation, Owner of Spirited Garden

真趣园也是一个拥有世界珍稀树种的私家盆景园。他们用罗汉松这种单一树种，以顽强的意志创造出世间独有的真趣松树种，这点让我很感动。

Zhenqu garden is also a precious place. They use the single species of *Podocarpus* to create works with a strong will, which makes me very moved.

【越南】阮氏皇 越南盆景协会主席

[Vietnam] Nguyen Thi Hoang the President of Vietnam Bonsai Association

关于黎德坚先生的真趣园：总体上它是一个宽敞秀美的盆景园，我可以看到主人在园中的投入。真趣园中有许多的树木藏品，特别值得一提的是，那里有许多高大的古树。然而，树木的种类却不是太多，它只专注于某种类型的植物。除此之外，它也忽略了某些细节。总而言之，真趣园适用于那些只是去享受大自然并且感受平静心灵的游客。

About Zhenqu garden: In general, it is a good and big garden. I can see that the owner invest so much in the garden. There are a big collection of trees in amount, and especially, there are many big and old trees. However, there are not much kinds of trees. It focus on only some type of plants. Besides, there are some details which are not taken care. It is suitable for visitor who just go to enjoy the nature, and feel the peace in the souls.

【美国】 威廉·尼古拉斯·瓦拉瓦尼斯 《国际盆栽》杂志出版人兼主编

[The United States] Valavanis Willima Nicholas Publisher and Editor for *International BONSAI* Magazine

东莞的真趣园令我感兴趣，因为我喜欢这里的罗汉松。之前我从没有看见过如此大又修剪那么好的罗汉松。另一个令我尤其感兴趣的是他们研发的新品种——真趣松，每年雨水最多的时候吐芽，每一次发芽时都会由鹅黄色慢慢变成红色，再变成浅绿，最后变成绿色，整个过程如同彩虹一般非常漂亮。我希望在参观的时候领略到这美丽的景象。这里平静的湖水和大又宽敞的园林是陈列盆景的好地方，也是举办室外晚会完美的场所。

The Zhenqu Garden in Dongguan interested me because I am fond of Podocarpus. I have never seen such large Podocarpus trained before. Of particular interest to me was the new variety they introduced which has red foliage in spring. I only wish they were colorful during my visit so I could appreciate the reddish hue myself. The large and spacious garden with the quiet lake was a perfect setting to display all the Penjing as well as to host the delicious outdoor banquet.

【日本】须藤雨伯 景道家元二世

[Japan] Uhaku Sudo Keido Iemoto (headmaster) II

拜访三个盆景园之后我感受盆景文化的发展是以中国盆景历史为重，以各个园子为豪，盆景人为之专心致志不懈努力的。可以看到各个园主对盆景的兴趣和对盆景未来造型的追求。各自拥有各自的理念与哲学，正在阔步前进。我认为盆景和庭院具体体现了园主的理念，目标明确，十分优秀。但是，从我们日本人的视角来看，大陆的和日本人的世界观还是稍有不同的。

About the three Penjing gardens we have visited: After the visit, I feel that the development of Penjing culture focuses on the history of Chinese Penjing, Penjing practitioners have devoted their heart and soul and feels proud of their own gardens. We can see the garden owner's interest for Penjing and their pursuits for the future modeling of Penjing. Each of them has their own concept and philosophy, and marches forward with big strides. I think that Penjings and gardens have concretely expressed the owner's concept, which is very clear and excellent. However, from the view of Japanese, there still are slight differences between the world views in the mainland and Japan.

「日本」須藤雨伯 景道家元二世

いづれも中国盆景の歴史を大切と各園の誇りをもって盆景文化の発展に一生懸命精進していることが伺われた。各園がそれぞれ盆景に対してどのような興味と将来の姿を求めているかがはっきり見ることが出来た。それぞれに理念と哲学があり、邁進している過程を感じた。盆景も庭園もその園主の理念が具現化するわけであるから、目的・目標がはっきりしていて素晴らしいと思う。しかし、われわれ日本人から見ると、やはり大陸的であり日本人の世界観とは少々違いを感じる。

【西班牙】 安东尼奥·帕利亚斯 《当代盆栽》杂志执行主编

[Spain] Antonio Payeras Editor and Director for *Bonsai ACTUAL* Magazine

国家盆景大展之后我们一行来到黎德坚先生位于广东东莞的真趣园。在这里我看到美丽的罗汉松，这个品种在我的国家很少见，所以在这里看到那么多优良品质的罗汉松树种令我感到惊奇。

After China National Penjing Exhibition, in Zhenqu Garden in Dong Guan, Guangdong Province where I discover the beauty of the varieties of *Podocarpus*. This is a rare species in my country, so for me it was amazing to see the quantity and good quality contained in this garden.

【丹麦】汤米·尼尔森 EBA 欧洲盆景协会成员国丹麦盆景协会会长

[Denmark] Tommy Nielsen, the President of Danish Bonsai Association

真趣园也令人印象深刻，与真趣园工作人员的谈话以及观赏园中的盆景作品，让我们非常开心。真趣园的晚餐也令人很满意、花园中的流水令人印象深刻，许多盆景的质量也非常高。

总而言之，这是一个非常漂亮的花园。但是我可以用更多的时间来发现花园里面蕴藏的更漂亮的事物。让我感到非常高兴的是，我和真趣园的主人进行了交谈，也了解了许多有关花园的故事。

The Zhenqu Garden was also impressive but it was after my opinions not as nice to visit because of the TV. The TV of course had an influence on the visit, but it was nice to talk to some of the staff from the garden and also to look at some of the Penjing wish was very impressive.In my opinion the visit was more formal then all the other visits and the dinner in the garden was a little different then the other dinners we had.

All in all it was a nice garden but i could have used more time there to have a bigger impression on what the garden had inside it and it would have been nice to have met the owner and heard his story about the garden...The lake was though very impressive and many of the Penjing was of very high quality.

【法国】克里斯蒂安·弗内罗法国《气韵盆栽》杂志出版人

[France] Christian Fournereau, Chief Publication for Esprit Bonsaï Magazine

【法国】米歇尔·卡尔比昂法国《气韵盆栽》杂志主编

[France] Michèle Corbihan, *Chief Redactor for Esprit Bonsaï* Magazine

真趣园稍显不同，因为它是专门用于公众参观的。非常有趣的是我们看到了许多非常高大的树木，但是它们却仍然在盆景盆中制作并作为盆景展出。此外，我们还非常幸运地能再次欣赏到这样的收藏。罗汉松在欧洲是一种非常罕见的物种，因为它们无法在我们国家所处的纬度生长。同样值得一提的是中国盆景大师对树木的呈现方式。我们看到的树木非常不同于我们在欧洲可以看到的树木。

The Zhenqu garden was different because it was organized for public visits. It was interesting to see very big trees but still presented as Penjing, in pots. There again we were very lucky to be allowed to admire such a collection. The Podocarpus is a very rare species in Europe as they don't grow under our latitudes. It was also interesting to see the Chinese way of presenting the trees. The trees we saw were very much in contrast to what we can see in Europe.

【马来西亚】蔡国华 国际盆栽俱乐部会员 马来西亚盆景雅石协会会员

[Malaysia] Dato'Chua Kok Hwa Member of Bonsai Clubs International, Member of Malaysia Bonsai N Stone Association

感谢真趣园为来宾举办热情洋溢的欢迎晚宴。真趣园里的盆景藏品不仅大而且世间罕见，尤其是那里的真趣松。

Many thanks Zhenqu Garden for their warm welcome and dinner party. Zhenqu Garden has good collections of large Penjing, especial the pine.

【日本】镰田照士 《近代盆栽》月刊主编

[Japan] Terushi Kamada the Chief Editor of the monthly magazine *KINBON*

带领我们参观的趣怡园的吴成发先生和真趣园的黎德坚先生的盆景园的规模让我感到震撼。真趣园让我看到了罗汉松嫁接的技术之高超。

I was shocked by the scale of Penjing gardens of Mr. Wu Chengfa, who led us for the visit of Quyi Garden, and Mr. Li Dejian, who led us for the visit of Zhenqu Garden. And I have seen the superb skill of podocarpus macrophyllus grafting in Zhenqu Garden.

「日本」鎌田照士 『月刊 近代盆栽』の編集長

また、その後ご案内頂いた趣怡園の呉成発氏、真趣園の黎徳堅氏の盆栽庭園の規模にも驚かされた。真趣園では、槙の枝接ぎなどの技術の高さも見せて頂くことが出来た。

【瑞典】玛利亚·阿尔博尔莉思·罗斯伯格 EBA 成员国瑞典盆景协会理事成员

[Sweden] Maria Arborelius-Rosberg the President of Swedish Bonsai Association

真趣园的参观也非常的精彩。在园中品尝美味的晚宴时，还欣赏到了精彩的表演。园中高挂的红灯笼，使得夜晚的景色更加的迷人。园主黎德坚先生也提供了精美的纪念品，酒店住宿也特别的棒。如果能有更多的时间去探索东莞就更好了。我想除了我之外，好多人都想有更多的自由时间参观东莞这座城市。当然，没有那么多的时间。对于我来说，所有的一切都是那么有趣，这是一个陌生的城市，有许多漂亮的地方值得探索。在这个城市，我看到的每一处都是那么的整洁，所有的植物和树木看上去都被悉心照料着。

Also the visit of the Zhenqu garden was very nice. We were very well served in the dinner party. Also we enjoyed the show very much. It was very beautiful at night with all the red lanterns! The lord gave us very fine gifts too. The hotel was very nice. It would have been nice to have a little more time to explore the city of Dongguan. I think more than me would have liked to have a little free time to go and visit the nice cities also, maybe with a guided tour. But, of course it was not time. For me all is so interesting because it is a new country with a lot of beautiful places to explore. I was very impressed how clean and nice the cities we saw were, all with a lot of fine plants and trees that looked well cared for.

【匈牙利】阿提拉·鲍曼 EBA(欧洲盆景协会)成员国匈牙利盆景协会副会长

[Hungary] Attila Baumann Vice President of Hungarian Bonsai Association

东莞的真趣园，在整体布局和风格方面有自己独特的风格。真趣园建成时间大约10年，园子中间有一个非常漂亮的湖，我觉得非常有趣；在园子里又有一个园子，入口看起来很逼真。没人会想到，真趣园也是真趣松的研究基地。我从来没有在同一个地方看到过这么多罗汉松。在真趣园用晚餐的时候还有十分精彩的表演，但是“变脸”这门艺术对于我们还是很陌生的，如果能事先介绍一些变脸的背景知识就更好了。

The Zhenqu Garden in Dongguan has its own characteristics in layout, structure and style. This garden was built in about 10 years with a beautiful lake in the center, and I found it very interesting; a garden was created in the garden itself, and the entrance of this looks very authentic. Nobody knew the fact, that this garden is also works as a research Centre for the Zhenqu pine. I have never seen so many Podocarpus in the same place. The dinner with the performance was amazing, however the performance of the masked woman was not clear to us. Some background information for foreigners would be helpful.

“琼楼玉宇” 玛瑙 长 50cm 宽 40cm 高 39cm 马建中藏品
“Richly Palace”. Agate. Length: 50cm, Width: 40cm, Height: 39cm. Collector: Ma Jianzhong

“红河艳” 大化石 长 27cm 宽 31cm 高 18cm 李正银藏品
“ Colourful Red Rival”. Macrofossil. Length: 27cm, Width: 31cm, Height: 18cm. Collector: Li Zhengyin

“瑞兽”大化玉石 长 60cm 宽 18cm 高 46cm 马建中藏品

“海洋之星” 大化玉石 长 30cm 宽 18cm 高 11cm 马建中藏品

“天上人间” 大化玉石 长 32cm 宽 28cm 高 17cm 马建中藏品

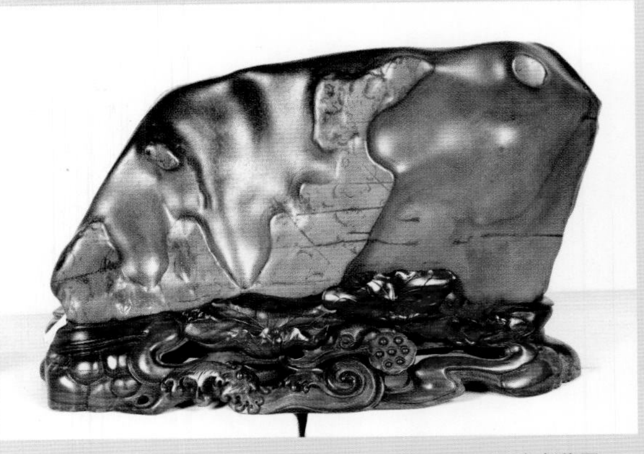

“海的女儿”彩陶石 长 60cm 宽 23cm 高 39cm 马建中藏品

2013 中国盆景会员展之赏石欣赏

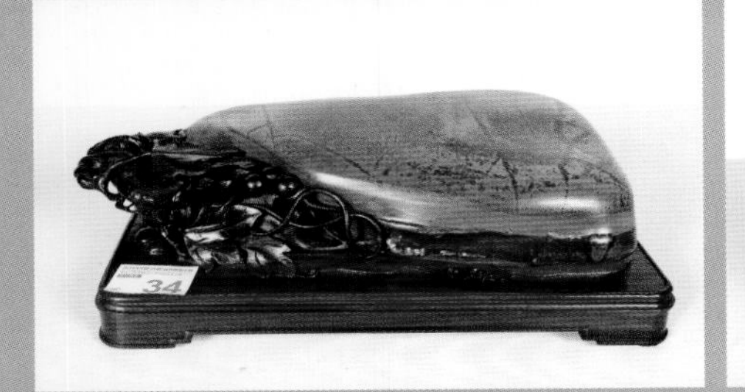

“硕果”大化玉石 长 42cm 宽 27cm 高 13cm 马建中藏品

“纯酿”大化玉石 长 25cm 宽 14cm 高 25cm 马建中藏品

“大树”玛瑙沙漠漆 长 25cm 宽 25cm 高 35cm 马建中藏品

“春色”蓝木化石 长 30cm 宽 20cm 高 25cm 马建中藏品

“俏佳人”玛瑙沙漠漆 长 14cm 宽 15cm 高 9cm 马建中藏品

“仙果”玛瑙沙漠漆 长 13cm 宽 16cm 高 9cm 马建中藏品

“神峰”玛瑙沙漠漆 长 48cm 宽 30cm 高 30cm 马建中藏品

“点将台”大化玉石 长 60cm 宽 35cm 高 26cm 马建中藏品

"雄狮"沙漠漆 长 30cm 宽 15cm 高 20cm 马建中藏品

"田螺"大化玉石 长 37cm 宽 26cm 高 27cm 马建中藏品

"双峰"硅化木石(内蒙)长 58cm 宽 32cm 高 65cm 马建中藏品

"仙居"沙漠漆 长 40cm 宽 36cm 高 64cm 马建中藏品

Scholar's Rocks

Appreciation of 2013China Penjing MemberExhibition

"黄金峰" 大化玉石 长 30cm 宽 26cm 高 20cm 马建中藏品

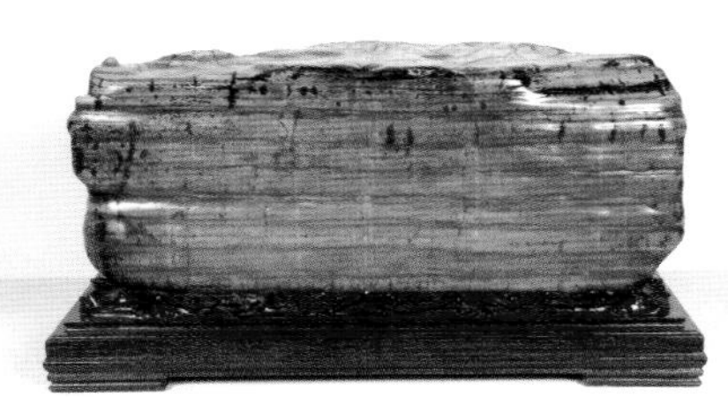
"平川之秋"大化石 长 65cm 宽 35cm 高 45cm 马建中藏品

"江南晚秋"大化玉石 长 32cm 宽 25cm 高 17cm 马建中藏品

"中国画"大化玉石 长 20cm 宽 9cm 高 37cm 马建中藏品

"高山流水"大化玉石 长 18cm 宽 10cm 高 45cm 马建中藏品

"琼楼玉宇"玛瑙 长 50cm 宽 40cm 高 39cm 马建中藏品

"美丽梯田" 大化玉石 长 42cm 宽 34cm 高 20cm 马建中藏品

"金枫"大化石 长 50cm 宽 20cm 高 40cm 马建中藏品

"神州大地" 大化玉石 长 65cm 宽 33cm 高 17cm 马建中藏品

"万水千山总是情"带沙漠漆的木化石 长 60cm 宽 40cm 高 64cm 马建中藏品

"天书" 大化石 长 67cm 宽 48cm 高 21cm 马建中藏品

"秋实" 大化石 长 46cm 宽 33cm 高 18cm 李正银藏品

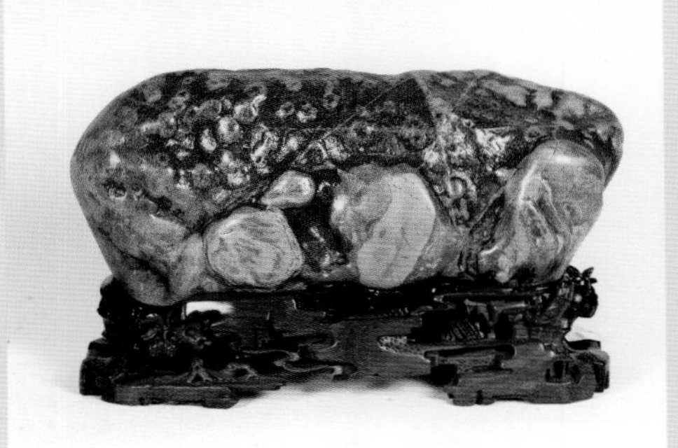
"米芾拜石" 乌江石 长 50cm 宽 25cm 高 12cm 李正银藏品

2013 中国盆景会员展之赏石欣赏

"凤栖梧桐" 乌江石 长 55cm 宽 33cm 高 17cm 李正银藏品

"如意" 大化石（左）长 23cm 宽 12cm 高 7cm（右）长 24cm 宽 12cm 高 9cm 李正银藏品

"千面人" 三江紫卵石 长 45cm 宽 28cm 高 19cm 李正银藏品

"至尊" 大化彩玉石 长 22cm 宽 20cm 高 18cm 李正银藏品

"金豆" 乌江石 长 40cm 宽 24cm 高 16cm 李正银藏品

"宝镜" 大化彩玉石 长 22cm 宽 29cm 高 9cm 李正银藏品

"观音" 乌江石 长 53cm 宽 25cm 高 22cm 李正银藏品

"北京猿人" 红彩陶石 长 54cm 宽 36cm 高 26cm 李正银藏品

"登云" 大化彩玉石 长 45cm 宽 30cm 高 10cm 李正银藏品

"龙首" 彩陶石 长 38cm 宽 33cm 高 19cm 李正银藏品

“红河艳”大化石 长 27cm 宽 31cm 高 18cm 李正银藏品

“灵崖仙境”大化石 长 38cm 宽 25cm 高 23cm 李正银藏品

“青蛙王子”乌江石 长 38cm 宽 32cm 高 26cm 李正银藏品

“高台胜景”大化石 长 30cm 宽 22cm 高 13cm 李正银藏品

Scholar's Rocks

Appreciation of 2013China Penjing MemberExhibition

“马踏飞燕”红彩卵石 长 41cm 宽 36cm 高 22cm 李正银藏品

“平台秋色”彩陶石 长 38cm 宽 20cm 高 16cm 李正银藏品

“龙鳞”大化石 长 38cm 宽 22cm 高 9cm 李正银藏品

“龙根”乌江石 长 36cm 宽 19cm 高 19cm 李正银藏品

“吉祥如意”大化石 长 36cm 宽 22cm 高 21cm 李正银收藏

“翡翠白菜”乌江石 长 40cm 宽 28cm 高 12cm 麻彩江收藏

“金顶”大化石 长 38cm 宽 28cm 高 22cm 李正银藏品

“金牛”大化石 长 43cm 宽 23cm 高 16cm 李正银藏品

“貔貅”大化石 长 46cm 宽 26cm 高 16cm 李正银藏品

韩国盆栽园集锦

Introduction of Korean Bonsai Garden

Halsurok盆栽园园主张星钧与心爱的景观树合影

纳米植物园园主金光水

中国盆景艺术家协会会长苏放与松树研究所园主李长熙合影

Saesol盆栽园园主李相煜与喜欢的盆景合影

清风盆栽教室园主赵明基在园子里合影

清州盆栽园园主李康熙

Gomsol盆栽研究院园主金大赫

德松山房园主金正太

木惜园园主全恩灿

仙游园园主金世元

Halsurok盆栽园 园主：张星钧
Halsurok Bonsai Garden Owner: Jang SeongGyun

张星钧从事盆栽35年，Halsurok盆栽园是韩国最大的盆栽批发基地，以盆栽、庭院树和野生花卉为主。

Bonsai history is 35 years. The biggest wholesale nursery in Korea who selling Bonsai, garden tree and wild flower.

纳米植物园 园主：金光水
Nami Plant Garden Owner: Kim Gwang Su

金光水从事盆栽25年，主要从事盆栽和庭院树配置。

Bonsai history is 25 years; his business is Bonsai and garden tree's distribution.

松树研究所 园主：李长熙
Pine Research Institution Owner: Lee Jang Hee

Saesol盆栽园 园主：李相煜
Saesol Bonsai Garden Owner: Lee Shang Wook

李相煜从事盆栽20年，擅长落叶树种的制作技艺，鹅耳枥属植物是其园子的主要品种。

Bonsai history is 20 years; he has special technique for deciduous trees. His major species is Carpinus.

清风盆栽教室 园主：赵明基
Breeze Bonsai Classroom Owner: Cho Myeng Ki

赵明基从事盆栽32年，他擅长绑扎和嫁接，尤其擅长赤松和木瓜的制作。

Bonsai history is 32 years, special technique is wiring and grafting. Special to Japan red pine and Cydonia.

清州盆栽园 园主：李康熙
Chengju Bonsai Garden Owner: Lee Gang Hee

李康熙从事盆栽35年，擅长绑扎、空中压条和嫁接技艺。清州盆栽园主要有赤松、真柏和落叶树种。

Bonsai history is 35 years, special technique is wiring, air layering, grafting. And special to Japan red pine, Juniperus and deciduous tree.

Gomsol盆栽研究院 园主：金大赫
Gomsol Bonsai Institution Owner: Kim DaeHyuk

金大赫是韩国第一大盆栽园的第二代继承人。有海松和很多不同形态的盆栽，以附石盆栽及露根盆栽为主。

He is the second generation, the biggest growing nursery in Korea, which are specially growing well Korean pine and many different styles. He has special technique to make and grow root over rock and exposed root style.

德松山房园主：金正太

Deoksongsanbang Owner: Kim Jeong Tae

金正太，盆栽爱好者，从事盆栽35年，他收藏坯材并亲自制作盆栽和庭院树。他擅长木瓜的制作技艺，且他的木瓜收藏世界第一！

Hobbyist, bonsai history is 35 years; he collected materials himself and made all Bonsai and garden tree by himself. He has special technique for Cydonia sinensis, and his Cydonia sinensis collection will be the best of the entire world!

木惜园园主：全恩灿

Mokseok Garden Owner: JeonEun Chan

全恩灿从事盆栽30年，其主要树种是赤松和庭院树，他擅长植物的快速繁殖。

Bonsai history is 30 yeras. His major species are red pine Bonsai and garden tree; he has special technique to grow fast.

仙游园园主：金世元

Seonyoo Garden Owner: Kim Sae Won

金世元是仙游园的第二代继承人。仙游园是韩国最大的盆栽出口基地，自1989年开始，盆栽已被出口至欧洲、美国、加拿大、亚洲等地。

Kim Sae Won is the second generation. The biggest Bonsai export nursery in Korea which exported Bonsai from 1989. Bonsai can be export from Korea to Europe, U.S, Canada, and Asia.

道峰农园园主：郑二锡

Daofeng Garden Owner: Jeong I Seok

郑二锡是韩国盆栽收藏第一人。

Jeong I Seok is the No1 bonsai collector in Korea.

赤松园园主：许允行

RedPine Garden Owner: Xu Yongxing

Halsurok盆栽园里朱木盆景壮硕的根部

韩国盆栽园简介 Introduction
Korean Bonsai Garden

纳米盆景园里的赤松盆景

Halsurok盆栽园户外的景观树直参云霄

Halsurok盆栽园别致的“盆景”大门

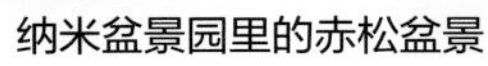

纳米植物园里的枫树盆景

道峰农园里的盆景

Saesol盆栽园中主干蜿蜒的盆景

Saesol盆栽园种植区的盆景

德松山房盆景种植区的景致

德松山房的盆景也像园子的名字一样有气势

德松山房盆景种植区一角

清风盆栽教室的盆景

仙游园里的枫树微型盆景

德松山房一角

德松山房户外种植的景观树也别有一番味道

Gomsol盆栽研究院里的景色

Gomsol盆栽研究院里根系养护细节欣赏

Gomsol盆栽研究院里错综盘绕的盆景树根

韩国盆栽园简介 Introduction Korean Bonsai Garden

木惜园户外种植的景观树

木惜园户外到处都是斜长的景观树

道峰农园里的这盆珍品也别有韵味

木惜园一角

仙游园微型盆景种植区的景色

道峰农园里盆景深处还藏着这样一件佳品

韩国盆栽园简介 Introduction
Korean Bonsai Garden

韩国第一收藏家、道峰农园园主郑二锡收藏的盆景

道峰农园角落里的盆景也是这样的精致

赤松园里绿意盎然

赤松园婀娜多姿的景观树不占少数

赤松园这棵树给人一种飘逸、宁静的感觉

赤松园一角

这棵双干景观树藏于赤松园地势低矮处非常吸引人的眼球

仙游园微型盆景种植区一角

施工中的开幕式舞台

加入中国传统文化视觉符号青花瓷元素的开幕式背景板，配合古朴典雅的清明上河图和富春山居图的背景墙，传达出浓郁的中国古韵

陆续到达的送展货车在场馆有序地排队，等待卸货、送

中国鼎

——2013（古镇）中国盆景国家大展展场花絮

展馆入口处，登记小组的佘博强、李建等在有条不紊地进行展品登记、编号发放等工作

广东省盆景协会副秘书长何焯光负责的会务小组在展馆入口处发放作品入选奖奖金

登记组的叶锦豪正在为已入场的展品细心地贴上编号

会员精品展组负责人温雪明在指挥布展工作

国家大展布展负责人陆志伟在安排大展展品的布置

盆景作品在入场前都经过精心的清洁工作

盆景作品在入场前都经过精心的清洁工作

China Ding
-2013(Guzhen) China National Penjing Exhibition
A Complete Collection of the Exhibition Scene

盆景作品在入场前都经过精心的清洁工作

盆景作品在入场前都经过精心的清洁工作

布展工作人员正小心翼翼地将一盆大型盆景放上展台

忙碌而有序的布展现场

工作人员在安装展品照明灯具

名贵的紫砂古盆已到位，正在准备布置

国家大展的 100 盆展品首次使用了独立展台，并加入了现代又别致的桌旗，在色彩搭配和造型设计上都给人以强烈的视觉冲击力，让展馆充满了简约又不失传统的艺术气息

2013年9月26日早上，2013（古镇）中国盆景国家大展布展组的工作在中国盆景艺术家协会常务副会长、广东省盆景协会会长曾安昌和广东省盆景协会和中国盆景艺术家协会副秘书长王金荣的带领下、在广东省各地方盆景协会的大力支持下，按中国盆景艺术家协会的设计方案热火朝天地全面展开了。

中国盆景艺术家协会名誉副会长陆志伟负责100盆国家大展展品的装卸、布展工作；广东省盆景协会常务理事温雪明负责200盆会员精品展的装卸、布置工作；中国盆景艺术家协会副会长申洪良和常务理事金文明负责50件古盆展

中盆协常务副会长、国家大展筹备委员会执行主任曾安昌接受电视台记者采访

中山市南方绿博园有限公司总经理、国家大展筹备委员会执行主任李金成正在接受媒体采访

会务组将本次大展、会员展的获奖证书、入选证书分类、分发

会务组将本次大展的获奖证书、入类、分发

国盆景艺术家协会常务副会长杨贵生、中国盆
艺术家协会副会长申洪良、中国盆景艺术家协
副会长关山、常务理事金文明参展的紫砂古盆

奇石珍品的送展现场

会员精品展展区简洁、淡雅，回形花纹与隔断既营造出中国古典美的气场氛围又衬托突出盆景主题

展品的运输、装卸和布展；柳州赏石协会常务理事张万龙负责52件奇石展品的布展工作。在各小组高度专业的工作和默契的配合之下，9月27日晚前，所有的展品都在广东省盆景协会常务理事佘博强负责的登记组登记备案，顺利完成进场及调整布置等工作。

所有送展作品在入场布展前都在展馆外进行了清洁洗礼，以最清新整洁的姿态出现在国家大展的展台上。一场前所未有、精彩绝伦的盆景视觉大展慢慢拉开帷幕。

经过5天的欢聚，大展圆满地落下帷幕，刷新了盆景展览史的纪录，创造了又一个辉煌。

国家大展的 100 盆展品首次使用了独立展台，并加入了现代又别致的桌旗，在色彩搭配和造型设计上都给人以强烈的视觉冲击力，让展馆充满了简约又不失传统的艺术气息

、会员
证书分

撤展现场，首届中国盆景的视觉盛宴落下帷幕

展场公布的大展、会员精品展评分表

第一次筹备会议 时间：2013 年 7 月 1 日；地点：南方绿博园会议室

第四次筹备会议 时间：2013 年 9 月 17 日；地点：南方绿博园会议室

2013（古镇）中国盆景国家大展筹备会议与总结会议实录

“思则有备，有备无患”，事前完全细致的筹备是举办任何大型成功活动必不可少的因素，中国盆景艺术家协会、广东省盆景协会和南方绿博园历时一年多的筹备工作涉及盆景展会的方方面面最终才使得汇集到全中国最高水平的盆景收藏和震撼视听的 2013 中国盆景年度之夜最终惊艳地呈现在所有中外盆景人的眼前。

第六次筹备会议 古镇镇副镇长何新煌发言 时间：2013 年 9 月 27 日；地点：南方绿博园会议室

第六次筹备会议 中国盆景艺术家协会会长苏放发言 时间：2013 年 9 月 27 日；地点：南方绿博园会议室

第五次筹备会议 时间：2013年9月23日；地点：南方绿博园会议室

The Records of 2013(Guzhen) China National Penjing Exhibition Preparatory Meetings and Wrap-up Meeting

一、2013（古镇）中国盆景国家大展筹备委员会第一次工作会议于2013年7月1日在中山市南方绿博园办公楼二层会议室召开。中国盆景艺术家协会、广东省盆景协会、南方绿博园三方代表共同参加此次筹备会议。本次会议主要就以下问题交换了意见并提出了解决方案：

1. 讨论了首届盆景展的实施方案。首届中国盆景国家大展成立组委会，负责策划所有盆景展的工作，组委会下设筹备委员会负责筹办展会日常事务工作。筹委会还将下设8个工作负责会务接待、宣传、展览等方面。中国盆景艺术家协会、广东省盆景协会、南方绿博园抽调相关工作人员负责具体事宜。并决定将筹委会办公室设在南方绿博园。

2. 确定了中国盆景国家大展三方——中国盆景艺术家协会、广东省盆景协会、中山市南方绿博园的业务联系人。

3. 明确每次会议的记录人员，并定期召开筹备工作会议，将展会的工作简报、会议纪要向中国盆景艺术家协会通报。

4. 本届盆景展的300盆盆景的资料收集和动态跟踪工作由中国盆景艺术家协会工作人员负责并全程汇总到中国盆景艺术家协会为国家大展特别设立的评比小组，由评比组筛选送展展品。

二、在第一次筹备会的基础上，2013年7月29日，筹委会召开了第二次会议，此次会议的重点放在2013（古镇）中国盆景国家大展的媒体宣传工作上，并细化了筹委组下设各小组的工作方案：

秘书组于8月3日前落实开幕式及晚宴的嘉宾名单并交付会务组。

于8月10日前完成展场设计方案的草稿，并于8月20日通过中国盆景艺术家协会、南方绿博园股份有限公司和广东省盆景协会三方商议最终定稿。

中国盆景艺术家协会、广东省盆景协会于8月15日前将参加本次展会的盆景作品名单（尺寸）提交到会务组，8月20日前落实所有参展作品。

会务组在8月30日前做好关于开幕式宴会的席位编排。

中国盆景艺术家协会于8月30日完成中国盆景年度之夜大会的方案。

所有的与会人员在会后实地视察本届盆景展的展览场地，并与展览公司现场进行展馆的布局调整。

三、在距2013（古镇）中国盆景国家大展开幕不到一个月的时间，2013年8月23日，筹委会召开了第三次会议，此次会议对展场的设计平面图、展台的设计风格以及嘉宾落实等情况进行了更加细节的讨论：

1. 讨论展览的布展平面图，最终确定选取10000m^{2}的展馆作为今次展览的场地，同时还讨论确定国家大展的展台风格、尺寸。

2. 落实受邀请嘉宾是否参会，中外嘉宾如有临时增减的情况再临时提

供名单。

3. 讨论嘉宾指南中的接待流程，车辆的安排。

四、2013 年 9 月 17 日，2013 年 9 月 23 日又连续召开了两次筹备会议，确保 2013（古镇）中国盆景国家大展能够万无一失地召开。

五、2013 年 9 月 27 日，2013（古镇）中国盆景国家大展开幕前两天，所有筹委会人员悉数到场，紧张地筹划、演练大展召开前的各个细节。

古镇镇副镇长何新煌先生、中国盆景艺术家协会苏放会长作为此次会议的主要发言人，从大展展场的布展情况、2013 中国盆景之夜的各个环节一一落实，坚决杜绝任何偏差和失误！各个小组的负责人，轮流汇报了自己职责范围内工作的落实情况。在大展开幕之前，在所有筹委会工作人员的共同努力之下，所有的准备工作都已就绪，激动地期待着 2013（古镇）中国盆景国家大展的开幕。

"善始善终"，2013（古镇）中国盆景国家大展的完美落幕，给所有中国盆景人一个精彩难忘的回忆，但是在其过程中还是会有或多或少的小瑕疵，我们希望总结这些不足，从而可以使得下届中国盆景国家更上一层楼，缔造更多的奇迹。

2013 年 12 月 1 日，在古镇镇政府会议室召开了 2013（古镇）中国盆景国家大展总结会。古镇镇副镇长何新煌、以苏放会长为首的中国盆景艺术家协会代表、广东省盆景协会代表共同参与到此次总结大会中。此次大会的宗旨是探讨交流、讨论不足和归纳总结。筹委会下设的各个小组的代表负责人轮流发言，总结各自工作的不足以及今后国家大展可以借鉴的地方。

2013（古镇）中国盆景国家大展筹备委员会执行主任、广东省盆景协会会长曾安昌在会上发表讲话，表示此次 2013（古镇）中国盆景国家大展最后成功地圆满完成，在国内外引起了强烈的反响，十分震撼人心。这其中的原因是：第一，此次盆景展的定位好——首届国家大展，与其他盆景协会一做好几届的活动截然不同，因为定位定的准、定的好。所以震撼、震惊。第二，从活动自始至终，我们做了大力的宣传，舆论决定一切。从一开始定位 100 盆精品，这在中国盆景界作为国家性的大型活动，是空前的，反映出我们要精品，对未来中国盆景艺术家协会和中国盆景界有重大影响，为中国盆景艺术家协会两年一次的国家大展打下了基础。第三，国家大展首席大奖"中国鼎"的创意设计，从中国古代"鼎"就是至高无上的权利的代表，把这样一个元素沿用到盆景，这在艺术领域里也是一个绝佳的定位，这种传统概念的引申也使人觉得震撼。

2013 中国盆景年度之夜开创了中国盆景界的先河，如奥斯卡颁奖仪式般的风格，融合了灯光舞美、音乐、沙画、T 台走秀等各种艺术表现手法，打破了中国盆景界晚宴的游戏规则。

中国盆景艺术家协会会长苏放，用了6个字来总结大展之后的感触——

总结会议 中国盆景艺术家协会会长苏放发言 时间：2013 年 12 月 1 日；地点：古镇镇政府

总结会议 广东省盆景协会会长曾安昌发言 时间：2013 年 12 月 1 日；地点：古镇镇政府

总结会议 古镇镇副镇长何新煌发言 时间：2013 年 12 月 1 日；地点：古镇镇政府

“谢天，谢地，谢人”。短短的 6 个字，却最贴切的表达了苏会长对这次大展能够成功举办而付出辛苦的人们的谢意。“谢天”——感谢全中国热爱盆景的人；“谢地”——感谢古镇政府，没有古镇政府的支持就不会有这次活动；“谢人”——感谢所有的工作人员。

何新煌镇长最后一个发言“盆景展落下帷幕不到两个月，中山市市领导以及党委书记也给予了充分的肯定，感谢中国盆景艺术家协会和广东省盆景协会给予的在座的所有工作人员表示感谢，通过这一次的总结可以为以后提供更多的借鉴！期待下次有更好的合作！”

2013（古镇）中国盆景国家大展在此次总结会后也画上了圆满的句号，让我们期待下一次的国家大展有更多的创举！

总结会议 时间：2013 年 12 月 1 日；地点：古镇镇政府

"中国罗汉研究示范

把享有罗汉松皇后美誉的"贵妃"罗汉松接穗嫁接到其他快速生长的罗汉松砧木上，生长速度比原生树还快几倍，亲和力强，两年后便能造型上盆观赏，这种盆景的快速成型的技术革命是谁完成的？是在哪里完成的？

THE RESURRECTION OF

CULTURE

BELONGING TO

TANG DYNASTY

STARTS FROM

HERE